CAROL VORDERMAN

HELP YOUR KIDS WITH

Times Tables

KEY STAGE 1 and 2

Written by Holly Beaumont, Joe Harris, Sean McArdle, Sue Phillips

Maths Consultant Sean McArdle

Asst. Editors Anamita Guha, Kritika Gupta

Editors Holly Beaumont, Rohini Deb, Joe Harris, Nishtha Kapil, Arpita Nath

Senior Editors Cecile Landau, Deborah Lock

Project Art Editor Tanvi Nathyal

Art Editors Dheeraj Arora, Gemma Fletcher, Alison Gardner, Hedi Hunter, Rashika Kachroo, Rosie Levine, Yamini Panwar, Lauren Rosier

Managing Editor Soma B. Chowdhury

Creative Director Jane Bull

Category Publisher Mary Ling

First published in Great Britain in 2016
This edition published in Great Britain in 2017
by Dorling Kindersley Limited
80 Strand, London WC2R 0RL

Material in this publication was previously published in
Easy Peasy Times Tables (2011); *Complete Times Tables Book* (2009);
10 Minutes a Day Times Tables (2014);
Maths Made Easy (Key Stage 1: Ages 5–7 Times Tables, 2014);
Maths Made Easy (Key Stage 2: Ages 7–11 Times Tables, 2014)

A CIP catalogue record for this book is available from the British Library.
ISBN: 978-0-2413-1701-3

Printed and bound in China

A WORLD OF IDEAS
SEE ALL THERE IS TO KNOW

CONTENTS

A NOTE FOR GROWN-UPS

There's a lot that you can do to get involved and help your child with their times tables. Look out for the tips for grown-ups dotted throughout this book. They include a whole host of suggestions about ways to introduce times tables in the context of everyday life, and how to employ a range of multi-sensory learning techniques. Building maths skills will build confidence and help your child to become a real maths whizz!

EASY PEASY

Times Tables

Consultant Sean McArdle

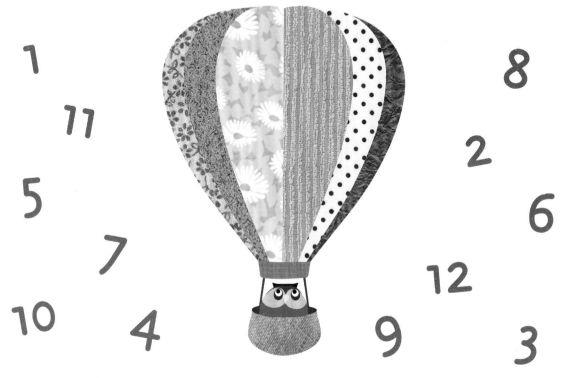

1
8
11
2
5
6
7
12
10
4
9
3

What are numbers?

Numbers are what we use to count things.
We can count all sorts of things...

Animals

People

Places

Objects

Units of measurement

Numbers can either be written as words or as symbols called numerals.

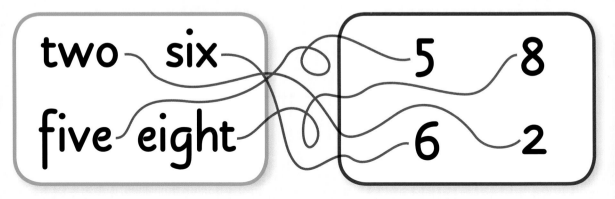

two   six

five   eight

5   8

6   2

Counting

Let's start by counting the objects below.

1 2 3 4 5 6 7 8 9 10

Tip for grown-ups

It's easier to understand numbers using counters that can be touched and arranged. Try making your own counters from objects like buttons, toy bricks, or pasta shapes.

7

Multiplication

It's easy to count small numbers of things.
But what if you have lots of things to count?

If we have lots of things to count,
it can be easier to count in groups.

We can count the socks in pairs.
There are 2 socks in every pair.

A jumble of socks.

$$2 + 2 + 2 + 2 + 2 = 10 \text{ socks}$$

This sum uses ADDITION.

How many pairs do we have? We have 5 pairs of socks.
This sum can also be written as

$$5 \times 2 = 10 \text{ socks}$$

This time we're using MULTIPLICATION.

5×2 is the same as 2×5.
The sums are written in different
ways but their answers are the same.

Another way of saying "multiplied by" is to say "TIMES".
The multiplication, or times, sign is ✕.

How many groups of apples do we have in these fruit bowls?

3 lots of 3 apples.

3 **+** 3 **+** 3

3 × 3 = 9 apples in total.

How many fish are in these fish bowls?

2 lots of 4 fish.

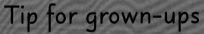

Multiplication is a quick way of adding up the same number over and over again.

4 **+** 4

2 × 4 = 8 fish in total.

Can you match up these addition and multiplication sums?

4 × 7

7 × 4

3 + 3 + 3

4 + 4 + 4 + 4 + 4 + 4 + 4

7 + 7 + 7 + 7

3 × 3

Tip for grown-ups

Adding and multiplying will be easier to understand if you start by drawing the number groupings or using homemade counters.

9

Times tables

Multiplication is quick and easy if we know our times tables. Times tables aren't pieces of furniture. They are mathematical tables that show us multiplication sums and answers.

The 1 times table is super easy.

$1 \times 1 = 1$

$2 \times 1 = 2$

$3 \times 1 = 3$

$4 \times 1 = 4$

$5 \times 1 = 5$

$6 \times 1 = 6$

$7 \times 1 = 7$

$8 \times 1 = 8$

$9 \times 1 = 9$

$10 \times 1 = 10$

$11 \times 1 = 11$

$12 \times 1 = 12$

"1 times" means having 1 group of something.

1 bowl with 3 apples. How many apples? **3**

1 bag of 7 sweets. How many sweets? **7**

1 dog with 4 feet. How many feet? **4**

Number patterns

The other times tables will be easier to remember if you get to know their number patterns. What number patterns can you see here?

Look at each number square and get to know the pattern for each times table.

1	2	3	4	5	6	7	8	9	10
11	12	13	14	15	16	17	18	19	20
21	22	23	24	25	26	27	28	29	30
31	32	33	34	35	36	37	38	39	40
41	42	43	44	45	46	47	48	49	50
51	52	53	54	55	56	57	58	59	60
61	62	63	64	65	66	67	68	69	70
71	72	73	74	75	76	77	78	79	80
81	82	83	84	85	86	87	88	89	90
91	92	93	94	95	96	97	98	99	100

These numbers are **odd** numbers.

These numbers all end in 0. They are all found in the 10 times table.

The numbers in white are **even** numbers. This is the pattern of the 2 times table.

Odd and even numbers

Numbers are either **odd numbers** or **even numbers**.

Odd numbers end in 1, 3, 5, 7, 9.
Even numbers end in 2, 4, 6, 8, 0.

Tip for grown-ups

This is a great time for colouring-in. Copy the number square out together and spend time colouring in the number pattern for each times table.

At the zoo with the twos

Lots of things come in twos – socks, shoes, hands, feet, eyes, and ears. Can you think of anything else?

How many penguins are there in this icy enclosure? They're standing in pairs so let's count them in 2s:

2   4   6   8   10

It's feeding time at the zoo. The zookeeper must feed the animals in the even-numbered pens. Which animals are these?

I'm hungry!

1

2

How many pairs of flamingos are there?

2×

Tip for grown-ups

Find opportunities to say the 2 times table together. Give it a try while pairing up socks, walking up the stairs, or passing houses in the street.

All the answers in the 2 × table are called "multiples" of 2.

1 × 2 = 2

2 × 2 = 4

3 × 2 = 6

4 × 2 = 8

5 × 2 = 10

6 × 2 = 12

7 × 2 = 14

8 × 2 = 16

9 × 2 = 18

10 × 2 = 20

11 × 2 = 22

12 × 2 = 24

2 × 6 = ___

How many legs do 6 flamingos have?

3 4 5 6

There are 3 pairs of flamingos.

13

Getting tiny with the tens

How many spots do the ladybirds have?

$3 \times 10 = 30$

The 10 times table is easy to learn. To multiply any whole number by 10, just add a 0 on the end.

2   3   6   8
4   5   7
1   9
10

Remember to add the 0.

$1 \times 10 = 1\,0$

$2 \times 10 = 2\,0$

$3 \times 10 = 3\,\_$

$4 \times 10 = \_\_\_$

Look, here are 10 ants walking in a row.

How many ants are there now?

$4 \times 10 = 40$

14   How many flies are zipping around?

Making numbers BIGGER

Let's make these numbers 10 times bigger! Remember, we do this by adding 0 to the end of each number.

×10

4 0

7 0

3 0

9 0

2 0

40 90 70 20 30

10×

1 × 10 = 10

2 × 10 = 20

3 × 10 = 30

4 × 10 = 40

5 × 10 = 50

6 × 10 = 60

7 × 10 = 70

8 × 10 = 80

9 × 10 = 90

10 × 10 = 100

11 × 10 = 110

12 × 10 = 120

How many seeds are there?

5 × 10 = ___

You should find 10.

In the sky with the fives

When you count in 5s, every other number ends in 5. All of the numbers in between end in a 0.

5 10 15 20 25 30 35 40 45 50...

There are 5 birds flying in this flock.

1 2 3 4 5

Numbers in the 5 times table end with either a 5 or a 0.

How many birds are flying in 4 flocks?

$$4 \times 5 = 20$$

How many birds are there in 5 flocks of 5?

The multiples of 5 run in two lines down the number square.

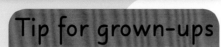

1	2	3	4	5	6	7	8	9	10
11	12	13	14	15	16	17	18	19	20
21	22	23	24	25	26	27	28	29	30
31	32	33	34	35	36	37	38	39	40
41	42	43	44	45	46	47	48	49	50
51	52	53	54	55	56	57	58	59	60
61	62	63	64	65	66	67	68	69	70
71	72	73	74	75	76	77	78	79	80
81	82	83	84	85	86	87	88	89	90
91	92	93	94	95	96	97	98	99	100

5×

$1 \times 5 = 5$

$2 \times 5 = 10$

$3 \times 5 = 15$

$4 \times 5 = 20$

$5 \times 5 = 25$

$6 \times 5 = 30$

$7 \times 5 = 35$

$8 \times 5 = 40$

$9 \times 5 = 45$

$10 \times 5 = 50$

$11 \times 5 = 55$

$12 \times 5 = 60$

Some bunches of balloons have drifted up into the sky. There are 5 balloons in each bunch.

Tip for grown-ups

You can use the five-day school week to work through the 5 times table together. How many school days are there in two weeks? Keep going until you get to 12 weeks.

There are _____ bunches of balloons.
There are _____ balloons in total.

25 birds.

Under the sea with the threes

How many birds are there?

$2 \times 3 = 6$

With a little help from under the sea, learning your 3 times table will be as easy as 1, 2, 3.

1. **3.** **2.**

These fish are swimming together in a group of 3.

Tip for grown-ups

Repeating the times tables out loud helps children to remember them. Making the 3 times table into a song or a sea shanty will make it even more memorable!

How many fish are there in 2 groups of 3?

How many crabs are scuttling along the sand?

$3 \times 3 = 9$

$2 \times 3 = 6$

How many fish are swimming in the sea?

3×

Number grid (1–100) with multiples of 3 circled:

1	2	**3**	4	5	**6**	7	8	**9**	10
11	**12**	13	14	**15**	16	17	**18**	19	20
21	22	23	**24**	25	26	**27**	28	**29**	**30**
31	32	**33**	34	35	**36**	37	38	**39**	40
41	**42**	43	44	**45**	46	47	**48**	49	50
51	52	53	**54**	55	56	**57**	58	59	**60**
61	62	**63**	64	65	**66**	67	68	**69**	70
71	**72**	73	74	**75**	76	77	**78**	79	80
81	82	83	**84**	85	86	**87**	88	89	**90**
91	92	**93**	94	95	**96**	97	98	**99**	100

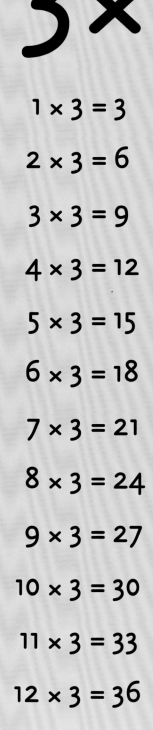

$1 \times 3 = 3$

$2 \times 3 = 6$

$3 \times 3 = 9$

$4 \times 3 = 12$

$5 \times 3 = 15$

$6 \times 3 = 18$

$7 \times 3 = 21$

$8 \times 3 = 24$

$9 \times 3 = 27$

$10 \times 3 = 30$

$11 \times 3 = 33$

$12 \times 3 = 36$

Because 3 is an odd number, its multiples are alternately odd and even numbers.

How many coins are piled up on the sand?

$5 \times 3 = \underline{\quad}$

You should find 15 fish in total.

19

In the garden with the fours

Let's try counting in 4s. Remember, 4 is an even number and all of its multiples will be even numbers too.

4   8   12   16   20   24...

Here are 4 plants. Each plant has one yellow flower. How many flowers are there? 4 of course!

How many red flowers are in the garden?

$2 \times 4 = 8$

How many snails are there in the garden?

Multiples of 4 end with the numbers 0, 2, 4, 6, or 8. Look, this pattern is repeated all through the square!

1	2	3	4	5	6	7	8	9	10
11	12	13	14	15	16	17	18	19	20
21	22	23	24	25	26	27	28	29	30
31	32	33	34	35	36	37	38	39	40
41	42	43	44	45	46	47	48	49	50
51	52	53	54	55	56	57	58	59	60
61	62	63	64	65	66	67	68	69	70
71	72	73	74	75	76	77	78	79	80
81	82	83	84	85	86	87	88	89	90
91	92	93	94	95	96	97	98	99	100

4×

$$1 \times 4 = 4$$

$$2 \times 4 = 8$$

$$3 \times 4 = 12$$

$$4 \times 4 = 16$$

$$5 \times 4 = 20$$

$$6 \times 4 = 24$$

$$7 \times 4 = 28$$

$$8 \times 4 = 32$$

$$9 \times 4 = 36$$

$$10 \times 4 = 40$$

$$11 \times 4 = 44$$

$$12 \times 4 = 48$$

If the rabbit takes a nibble out of every 4th carrot in the row, which carrots will he eat?

1 2 3 4 5 6 7 8 9 10 11 12

Tip for grown-ups

The morning is the best time to test and practise times tables – when brains are fresh and children are energized. Short bursts of activity are more effective than long sessions.

There are 12 snails.

Magic time with the nines

The 9 times table may seem a little difficult but the wise old owl can show you a trick or two.

That's magic! The owl has produced 9 white rabbits from his top hat.

1 2 3 4

5 6 7 8 9

How many rabbits has the hat produced this time?

2 × 9 = 18

How many rabbits in total?

As you go through the 9 times table, the first digit in the number goes up by 1 and the last digit goes down by 1 each time.

1	2	3	4	5	6	7	8	9	10
11	12	13	14	15	16	17	18	19	20
21	22	23	24	25	26	27	28	29	30
31	32	33	34	35	36	37	38	39	40
41	42	43	44	45	46	47	48	49	50
51	52	53	54	55	56	57	58	59	60
61	62	63	64	65	66	67	68	69	70
71	72	73	74	75	76	77	78	79	80
81	82	83	84	85	86	87	88	89	90
91	92	93	94	95	96	97	98	99	100

9×

$$1 \times 9 = 9$$

$$2 \times 9 = 18$$

$$3 \times 9 = 27$$

$$4 \times 9 = 36$$

$$5 \times 9 = 45$$

$$6 \times 9 = 54$$

$$7 \times 9 = 63$$

$$8 \times 9 = 72$$

$$9 \times 9 = 81$$

$$10 \times 9 = 90$$

$$11 \times 9 = 99$$

$$12 \times 9 = 108$$

The magic number

The number 9 and its multiples are special numbers. Each multiple of 9 can be broken down and added together to produce… the number 9!

$$2 \times 9 = 1 \quad 8$$

$$1 + 8 = 9$$

$$3 \times 9 = 2 \quad 7$$

$$2 + 7 = 9$$

Which of these numbers are multiples of 9?

56   27   93   54   9

32   15   73   44   81   72

At the sweet shop with the **sixes**

Life is sweeter when you know your times tables, and it's easier to work out prices and amounts in shops.

Here are 6 large lollipops, all bright and sugary.

Here are 2 groups of 6 lollipops. How many are there now?

$$2 \times 6 = 12$$

Remember, you can write the sum 2 × 6 or 6 × 2, the answer will be the same.

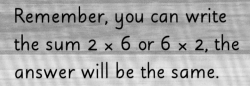

How many marshmallows are in these bags?

1	2	3	4	5	**6**	7	8	9	10
11	**12**	13	14	15	16	17	**18**	19	20
21	22	23	**24**	25	26	27	28	29	**30**
31	32	33	34	35	**36**	37	38	39	40
41	**42**	43	44	45	46	47	**48**	49	50
51	52	53	**54**	55	56	57	58	59	**60**
61	62	63	64	65	**66**	67	68	69	70
71	**72**	73	74	75	76	77	**78**	79	80
81	82	83	**84**	85	86	87	88	89	**90**
91	92	93	94	95	**96**	97	98	99	100

The multiples of 6 form rows running diagonally across the number square.

6×

$1 \times 6 = 6$

$2 \times 6 = 12$

$3 \times 6 = 18$

$4 \times 6 = 24$

$5 \times 6 = 30$

$6 \times 6 = 36$

$7 \times 6 = 42$

$8 \times 6 = 48$

$9 \times 6 = 54$

$10 \times 6 = 60$

$11 \times 6 = 66$

$12 \times 6 = 72$

Each jar on this shelf contains 6 sweets. How many sweets are there in total?

$5 \times 6 = \underline{}$

Tip for grown-ups

Time to raid the piggy bank! Practise multiplying with real money or when you're in a shop. This will show how useful the times tables can be.

There are 24 marshmallows.

In the heavens with the sevens

We're reaching the heights with our times tables now. Once you've learned your 7 times table, you'll be one step nearer to becoming a master of multiplication.

1 2 3 4 5 6 7

Here is a cluster of 7 stars.

How many stars are there in 3 clusters of 7?

$3 \times 7 = 21$

How many shooting stars can you see?

$2 \times 7 = \underline{\ \ }$

How many antennae do the Martians have?

1	2	3	4	5	6	**7**	8	9	10
11	12	13	**14**	15	16	17	18	19	20
21	22	23	24	25	26	27	**28**	29	30
31	32	33	34	**35**	36	37	38	39	40
41	**42**	43	44	45	46	47	48	**49**	50
51	52	53	54	55	**56**	57	58	59	60
61	62	**63**	64	65	66	67	68	69	**70**
71	72	73	74	75	76	**77**	78	79	80
81	82	83	**84**	85	86	87	88	89	90
91	92	93	94	95	96	97	**98**	99	100

7 ×

$1 \times 7 = 7$

$2 \times 7 = 14$

$3 \times 7 = 21$

$4 \times 7 = 28$

$5 \times 7 = 35$

$6 \times 7 = 42$

$7 \times 7 = 49$

$8 \times 7 = 56$

$9 \times 7 = 63$

$10 \times 7 = 70$

$11 \times 7 = 77$

$12 \times 7 = 84$

Multiplying Martians

These creatures are Martians. They live in outer space. Each Martian has:

3 antennae

4 eyes

5 arms

8 legs

How many eyes, legs, and arms do we have in total?

$7 \times 4 =$ ___ eyes

$7 \times 8 =$ ___ legs

$7 \times 5 =$ ___ arms

They have 21 antennae in total.

27

Baking cakes with the eights

The multiples of 8 are all even numbers.

The times tables come in really useful when we're baking and making things. They can help us to work out the right amounts.

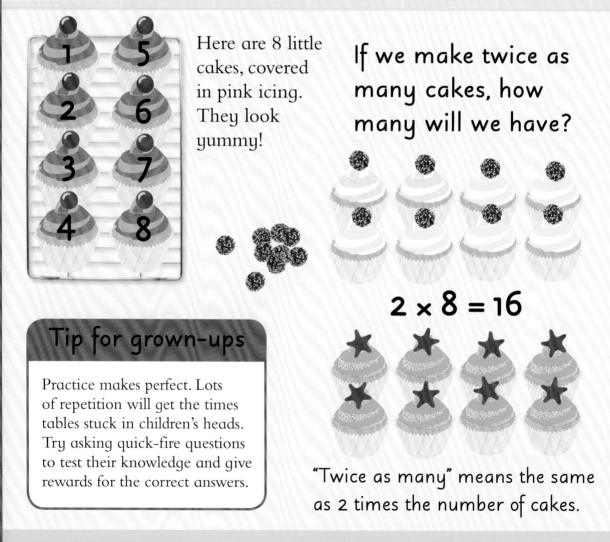

Here are 8 little cakes, covered in pink icing. They look yummy!

If we make twice as many cakes, how many will we have?

$2 \times 8 = 16$

"Twice as many" means the same as 2 times the number of cakes.

Tip for grown-ups

Practice makes perfect. Lots of repetition will get the times tables stuck in children's heads. Try asking quick-fire questions to test their knowledge and give rewards for the correct answers.

How many cherries are there?

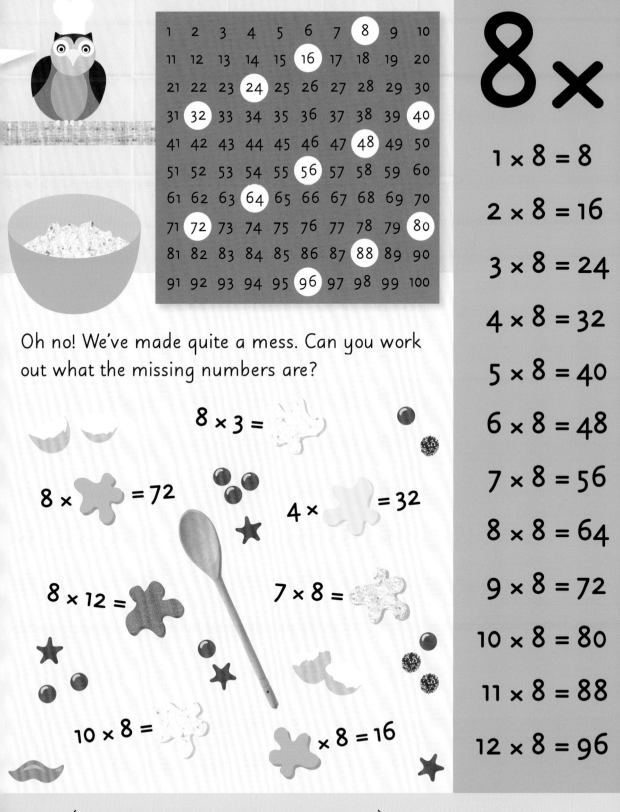

1	2	3	4	5	6	7	**8**	9	10
11	12	13	14	15	**16**	17	18	19	20
21	22	23	**24**	25	26	27	28	29	30
31	**32**	33	34	35	36	37	38	39	**40**
41	42	43	44	45	46	47	**48**	49	50
51	52	53	54	55	**56**	57	58	59	60
61	62	63	**64**	65	66	67	68	69	70
71	**72**	73	74	75	76	77	78	79	**80**
81	82	83	84	85	86	87	**88**	89	90
91	92	93	94	95	**96**	97	98	99	100

8 ×

$$1 \times 8 = 8$$

$$2 \times 8 = 16$$

$$3 \times 8 = 24$$

$$4 \times 8 = 32$$

$$5 \times 8 = 40$$

$$6 \times 8 = 48$$

$$7 \times 8 = 56$$

$$8 \times 8 = 64$$

$$9 \times 8 = 72$$

$$10 \times 8 = 80$$

$$11 \times 8 = 88$$

$$12 \times 8 = 96$$

Oh no! We've made quite a mess. Can you work out what the missing numbers are?

$$8 \times 3 =$$

$$8 \times = 72$$

$$4 \times = 32$$

$$8 \times 12 =$$

$$7 \times 8 =$$

$$10 \times 8 =$$

$$ \times 8 = 16$$

There are 16 (8 on the cakes and 8 loose).

Activities

Here are some fun puzzles to test your times tables knowledge.

Take your time solving these puzzles and practise any times tables you find it hard to remember.

Spaghetti tangle
Can you untangle this puzzle and find the matching pairs?

2 × 9		20
3 × 4		18
5 × 4		12
6 × 2		80
8 × 10		12

Give a dog a home
Can you work out which kennel belongs to each dog?

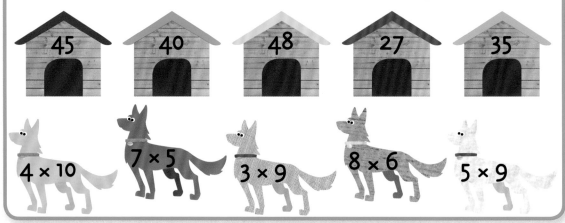

45   40   48   27   35

4 × 10   7 × 5   3 × 9   8 × 6   5 × 9

Going fishing

Match the number on each fish to the times table it appears in.

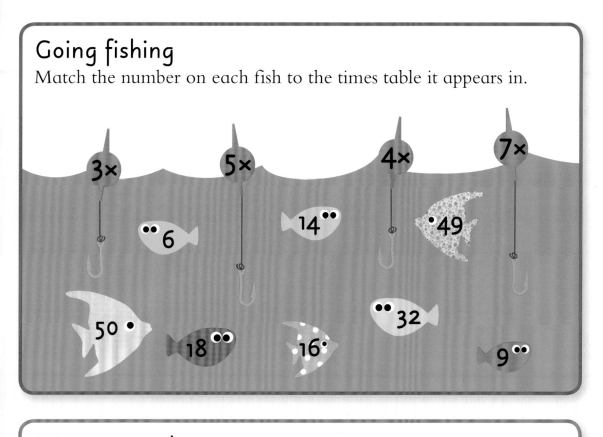

Missing numbers

Which person is needed to complete each multiplication?

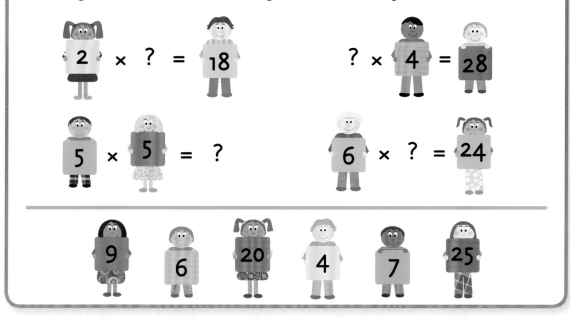

Activities answers

Spaghetti tangle

2 × 9 = → 18

3 × 4 = → 12

5 × 4 = → 20

6 × 2 = → 12

8 × 10 = → 80

Going fishing

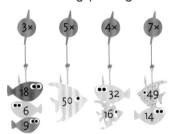

Give a dog a home

Missing numbers

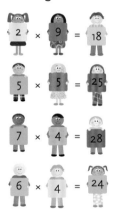

2 × 9 = 18

5 × 5 = 25

7 × 4 = 28

6 × 4 = 24

TRICKY

TiMES TaBLeS

3x 8x 10x 12x

5x 9x 2x 4x

Contents

Turn the page, and we'll get started.

HELLO!

Learn with games, puzzles, and magic.

The times tables can be fun!

3x

10x

7x

Introduction

The times tables are special shortcuts that make difficult maths fast and easy. They do this by telling you the answers to multiplication and division problems.

Multiplication

Multiplication is a fast way of adding up.

How many apples are there on these trees?
There are two ways you could find out.

The slow way: You could add together all the apples like this: **7 + 7 + 7 + 7 = 28**.

The speedy way: Or you could multiply them together, like this: **7 × 4 = 28**.

When you see these words, get ready for multiplication:

times   lots of
groups of
multiply   double

Division

Division is a fast way of subtracting until you approach or reach zero.

If you pick **15** apples, how many can go in each of these apple pies?
You need to find out how many lots of five are the same as **15**.

When you see these words, get ready for division:

share   between
divide
equal groups   into

The slow way: You could find out by subtracting, like this: **15 - 5 - 5 - 5 = 0**. That shows that **3** lots of **5** are the same as **15**, so **3** apples can go in each pie.

The speedy way: Or you could divide them together, like this: **15 ÷ 5 = 3**.

When do we use the times tables?

We use the times tables constantly in our everyday lives.

3 eggs × 3 = 9 eggs.

When we're cooking.

2 pounds × 3 = 6 pounds.

When we're shopping.

10 slices ÷ 5 = 2 slices each.

When we're sharing.

We use the times tables all the time – sometimes without knowing it!

2 points × 6 = 12 points.

When we're playing sports.

5 minutes × 5 = 25 minutes past.

When we're telling the time.

Learning with songs

It's easy to remember words when they're put to a tune. To help you learn your times tables, we've put together some songs that you can download from the Dorling Kindersley website.

The songs repeat each times table five times. Try to join in with singing the times tables as soon as you can. You can play a game by trying to call out the answers before they're sung. If you like, you can dance to the music too!

To download the songs, go online to this internet address: www.dk.com/timestables.

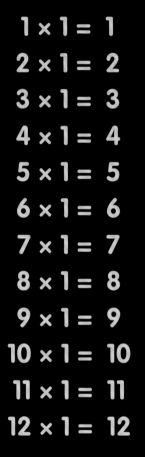

The one times table

When you multiply by one, the answer is the same as the number you started with. Nothing changes.

Here's the 1 times table:

$1 \times 1 = 1$

$2 \times 1 = 2$

$3 \times 1 = 3$

$4 \times 1 = 4$

$5 \times 1 = 5$

$6 \times 1 = 6$

$7 \times 1 = 7$

$8 \times 1 = 8$

$9 \times 1 = 9$

$10 \times 1 = 10$

$11 \times 1 = 11$

$12 \times 1 = 12$

Read the answers to the one times table. You will be counting from **1** to **12**.

Just one lot of…

Multiplying by one means the same as having one lot of something.

For example, **1** bag of **7** marbles. How many marbles? **7**.

1 bag of 7 marbles

How many?

1 net of **3** fish. How many fish?

1 flower with **6** petals. How many petals?

1 basket with **12** apples. How many apples?

1 purse with **7** coins. How many coins?

The one times mirror

Multiplying a number by **1** is like putting it in front of the mirror. You see the same thing again.

So, **8 × 1 = 8**.

The zero times table

When you multiply by zero, you're saying that there are zero lots of something. In other words, nothing at all.

Here's the 0 times table:

$1 \times 0 = 0$
$2 \times 0 = 0$
$3 \times 0 = 0$
$4 \times 0 = 0$
$5 \times 0 = 0$
$6 \times 0 = 0$
$7 \times 0 = 0$
$8 \times 0 = 0$
$9 \times 0 = 0$
$10 \times 0 = 0$
$11 \times 0 = 0$
$12 \times 0 = 0$

Nothing at all

If you know that something is empty, it doesn't matter how many lots of it you have.

For example: **1** sweets jar with **0** sweets. How many sweets? **0**.

1 jar with 0 sweets

2 birdcages with **0** birds. How many birds?

3 ponds with **0** frogs. How many frogs?

4 baskets with **0** eggs. How many eggs?

Calculator corner

1,000,000

Type 1,000,000 into your calculator, then press "× 0 =". What is the answer? This shows that it doesn't matter how big a number is – multiplying by zero still makes zero.

The ferocious number eater

The number **0** is like a ravenous monster. If you put any other number in a multiplication with **0**, the **0** will eat it up.

Answers: 0 birds, 0 frogs, 0 eggs.

The two times table

2×

The two times table is all about doubling, halving, and pairs. It's quick to learn, and easy to use.

Here's the 2 times table:

1 × 2 = 2
2 × 2 = 4
3 × 2 = 6
4 × 2 = 8
5 × 2 = 10
6 × 2 = 12
7 × 2 = 14
8 × 2 = 16
9 × 2 = 18
10 × 2 = 20
11 × 2 = 22
12 × 2 = 24

Counting in pairs

Many everyday things come in pairs. You can count them faster by counting in twos, like this: **2, 4, 6, 8, 10, 12, 14, 16, 18, 20, 22, 24**.

A pair of shoes

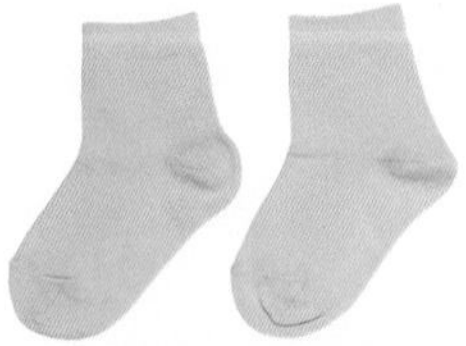
A pair of socks

A pair of gloves

Count these in groups of two

How many shoes in **3** pairs?

The answer is **6**.

How many socks in **5** pairs?

How many gloves in **6** pairs?

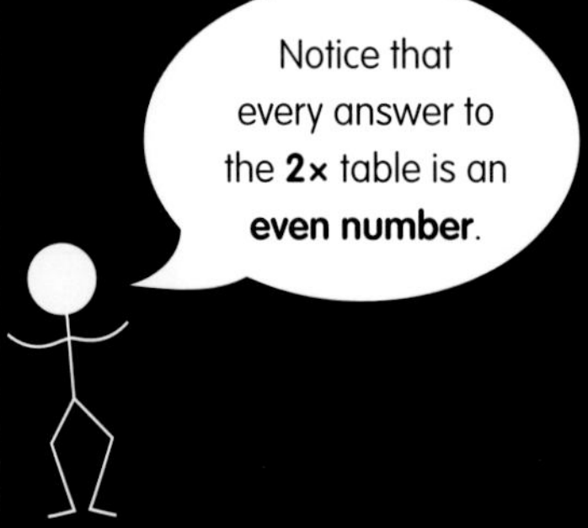
Notice that every answer to the 2× table is an even number.

Answers: 10 socks, 12 gloves.

Counting pairs is a kind of multiplication. Instead of writing "Four pairs are eight," you can write that "**4 x 2 = 8**." This is because a pair is a group of two.

4 x 2 = 8

Odd and even numbers

Even numbers end in

2 4 6 8 0

Odd numbers end in

1 3 5 7 9

Can you tell whether these numbers are odd or even?

52
436
452,789

All the answers in the **2x** table end in an even number. This pattern will help you to remember them.

1	**2**	**3**	**4**	**5**	**6**	**7**	**8**	**9**	**10**	**11**	**12**
odd	even	odd	even	odd	even	odd	even	odd	even	odd	even

Doubling machine

You can think of the **2x** table as an incredible doubling machine. Whatever you put in, twice as much comes out! Wouldn't it be handy to have a machine like that?

REMEMBER, REMEMBER

Read out loud all the questions and answers in the **2x** table. Then use your hand or a bookmark to cover up the answers, and try saying them again. Can you get them all right?

1 sock, 1 teddy bear, 2 footballs and 4 coins go in.

IN

DOUBLING MACHINE

How many footballs are there?

How many coins come out?

How many teddy bears are there?

How many socks come out?

OUT

Answers: 52 is even, 436 is even, 452,789 is odd, 4 footballs, 8 coins, 2 teddy bears, 2 socks.

2×

Let's go shopping

The **2×** table can help you to work out how much things cost. This will be very helpful when you go to the shops!

How many toffees can I afford?

SWEET SHOP

Two times toffees

Each of these toffees costs **2 pence**.   2p

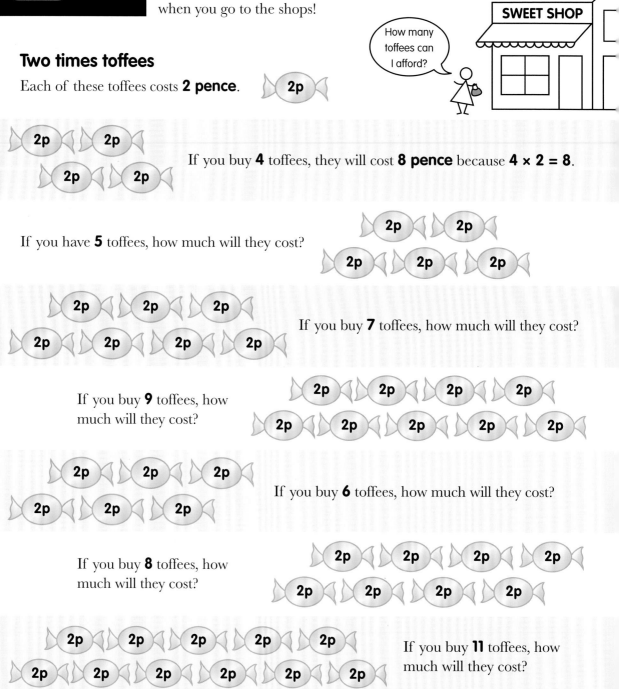

2p   2p
2p   2p

If you buy **4** toffees, they will cost **8 pence** because **4 × 2 = 8**.

If you have **5** toffees, how much will they cost?

2p   2p
2p   2p   2p

2p   2p   2p
2p   2p   2p   2p

If you buy **7** toffees, how much will they cost?

If you buy **9** toffees, how much will they cost?

2p   2p   2p   2p
2p   2p   2p   2p   2p

2p   2p   2p
2p   2p   2p

If you buy **6** toffees, how much will they cost?

If you buy **8** toffees, how much will they cost?

2p   2p   2p   2p
2p   2p   2p   2p

2p   2p   2p   2p   2p
2p   2p   2p   2p   2p   2p

If you buy **11** toffees, how much will they cost?

42

Be fair – share!

Imagine you and a friend are sharing out **24** toffees between you. How many toffees will you each have?

How many toffees × 2 = 24?

This is a bit trickier.

Calculator corner

$2 \times 2 =$

What do you think $2 \times 2 \times 2 \times 2$ equals? Try typing it into your calculator. How large a number do you think you will get if you type 2, then "× 2 =" 20 times? Get ready for a surprise.

Paperboy

You can think of the **2×** table as a number line. This boy is delivering newspapers. He drops them off at every other house. The houses that he stops at are the same as the answers to the two times table. Where will he stop next?

The word "multiple" comes from "multiplication". 6 is a multiple of 2, because 2 can be multiplied by another number (3) to make 6.

Odd ones out

Which of these numbers are not multiples of **2**? (Remember that all the answers in the **2×** table are even numbers.)

TOP TIP

If you know how to add, the **2×** table is easy. Just remember that two times a number means the same as adding that number to itself.

5 × 2 is the same as **5 + 5**.

Answers: 12 toffees × 2=24. The paperboy will stop at number 14 next. 7, 15, 9, 13, and 21 are not multiples of two.

5×

The five times table

All that you need to count in fives are your hands. And if you can count in fives, then you can multiply by five too!

Have you spotted the pattern?

Here's the 5 times table:

1 × 5 = 5
2 × 5 = 10
3 × 5 = 15
4 × 5 = 20
5 × 5 = 25
6 × 5 = 30
7 × 5 = 35
8 × 5 = 40
9 × 5 = 45
10 × 5 = 50
11 × 5 = 55
12 × 5 = 60

Yes, look – all the answers in the **5×** table end in **5** or **0**!

If you multiply **5** by an **odd** number, the answer will end with **5**.

If you multiply **5** by an **even** number, the answer will end with **0**.

Counting in fives

All these things come in groups of five. You can use the **5×** table to quickly add them up. Count like this: **5, 10, 15, 20, 25, 30, 35, 40, 45, 50, 55, 60**.

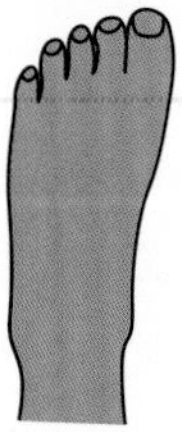

five toes

five arms

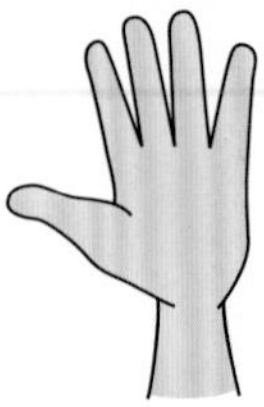

five fingers

five petals

Count these in groups of five

How many fingers on **4** hands?

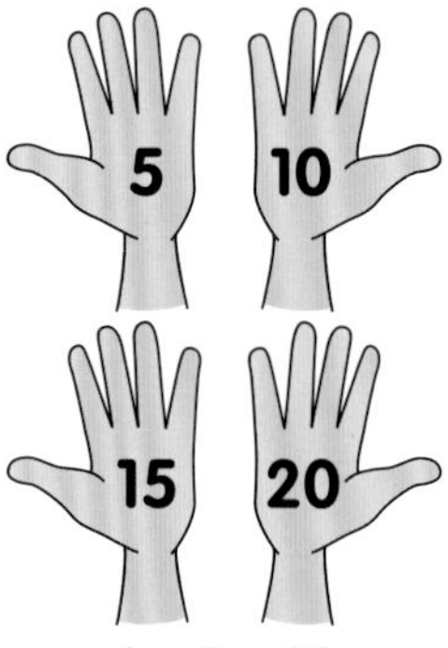

4 × 5 = 20

How many arms on **9** starfish?

How many petals on **7** flowers?

How many toes on **6** feet?

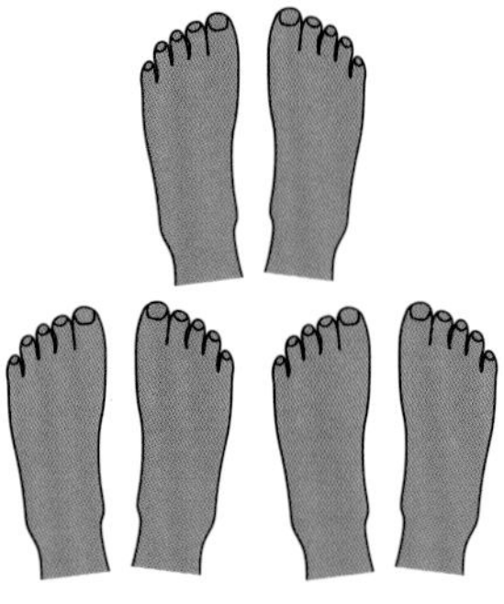

Answers: 45 arms, 35 petals, 30 toes.

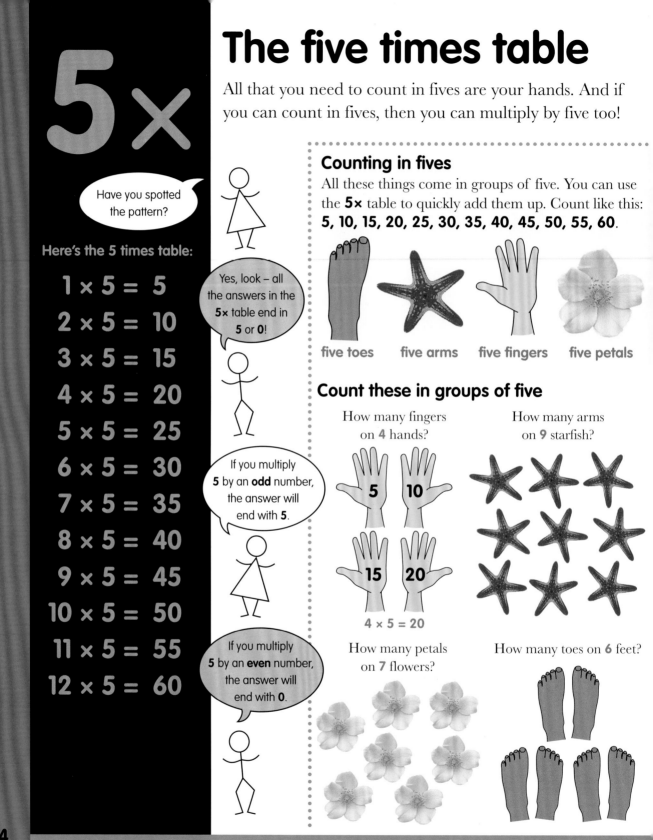

Practice makes perfect

Read the questions and answers in the **5×** table out loud. Then give this book to a friend, and ask them to test you. Can you remember the answers?

Backwards and forwards

Have you noticed that the answer to **2 × 5** is the same as the answer to **5 × 2**?

$$5 × 2 = 10$$
$$2 × 5 = 10$$

Five **2-pence** toffees cost exactly the same as two **5-pence** lollipops.

Multiplications give the same answer whichever way round you put the numbers. This means that you already know some of the answers to the other times tables.

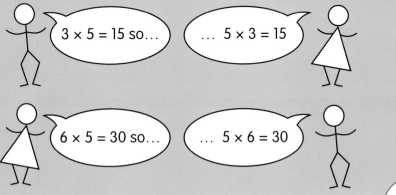

3 × 5 = 15 so... ... 5 × 3 = 15

6 × 5 = 30 so... ... 5 × 6 = 30

Cleaning windows

You can also think of the **5×** table as a number line. Imagine a window cleaner who stops at every fifth floor of a skyscraper. Where will he stop next?

Don't look down!

24	
23	
22	
21	
20	
19	
18	
17	
16	
15	
14	
13	
12	
11	
10	
9	
8	
7	
6	
5	
4	
3	
2	
1	

The cleaner's next stop will be the 25th floor.

45

5×

Five times clock

The **5×** table helps you to tell the time. There are **5** minutes between each number on the clock face.

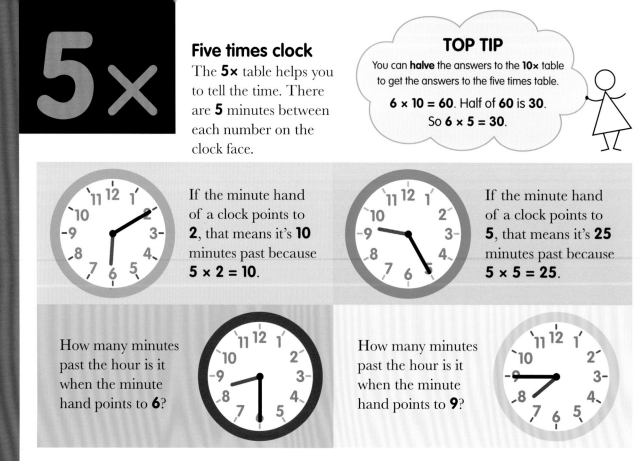

If the minute hand of a clock points to **2**, that means it's **10** minutes past because **5 × 2 = 10**.

If the minute hand of a clock points to **5**, that means it's **25** minutes past because **5 × 5 = 25**.

How many minutes past the hour is it when the minute hand points to **6**?

How many minutes past the hour is it when the minute hand points to **9**?

Rows and columns

You can use the times tables to count objects in rows and columns. First count how many columns there are, then count how many rows. Multiply those numbers together.

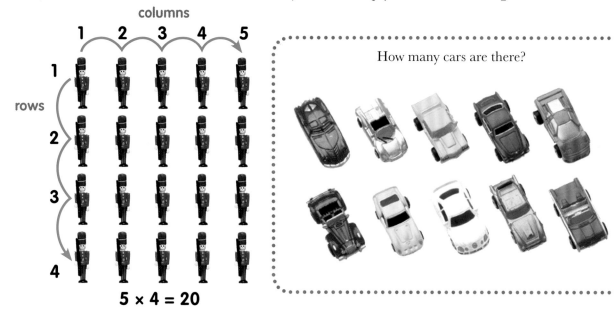

columns

1 2 3 4 5

rows

1

2

3

4

5 × 4 = 20

How many cars are there?

46

Let's go shopping again

Each of these coins is worth **5** pence.

5p

I hope I can afford that football.

TOY SHOP

5p
5p **5p**

If you have **3** coins, they are worth **15 pence** because **5 × 3 = 15**.

If you have **7** coins, how much are they worth?

5p **5p** **5p**
5p **5p** **5p** **5p**

5p **5p**
5p **5p**

If you have **4** coins, how much are they worth?

If you have **9** coins, how much are they worth?

5p **5p** **5p** **5p**
5p **5p** **5p** **5p** **5p**

5p **5p** **5p**
5p **5p** **5p**

If you have **6** coins, how much are they worth?

How many deckchairs are there?

How many buttons are there?

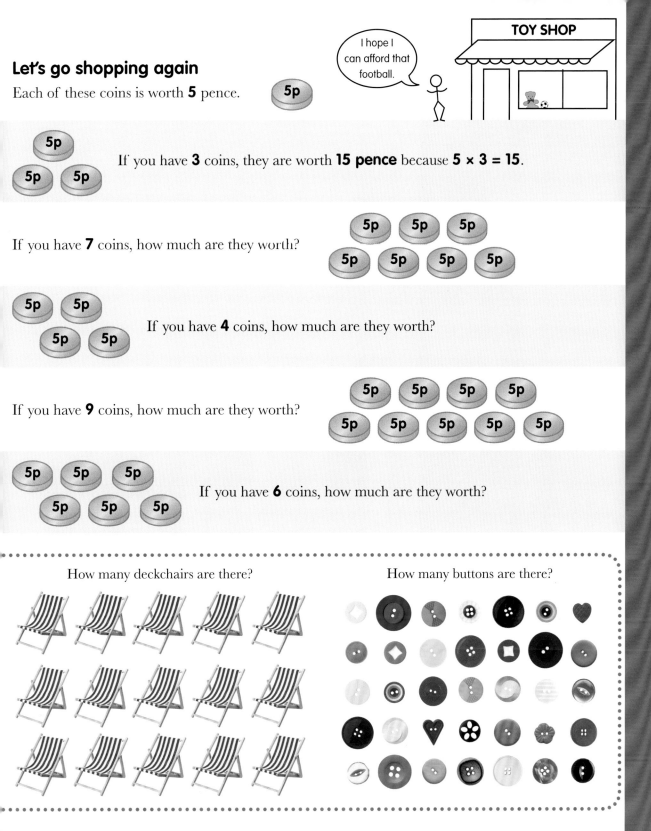

47

10×

Here's the 10 times table:

1 × 10 = 10
2 × 10 = 20
3 × 10 = 30
4 × 10 = 40
5 × 10 = 50
6 × 10 = 60
7 × 10 = 70
8 × 10 = 80
9 × 10 = 90
10 × 10 = 100
11 × 10 = 110
12 × 10 = 120

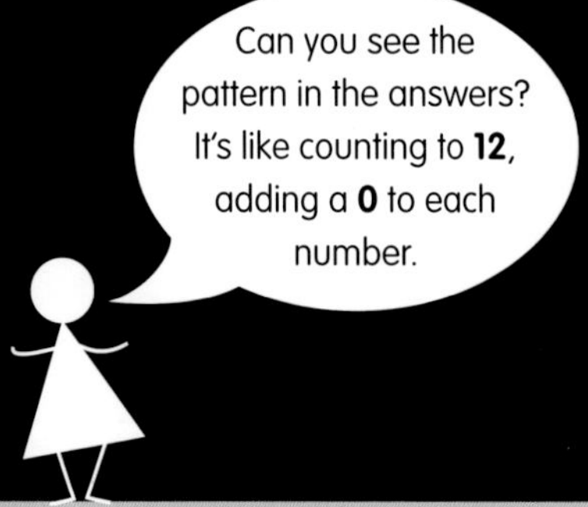
Can you see the pattern in the answers? It's like counting to **12**, adding a **0** to each number.

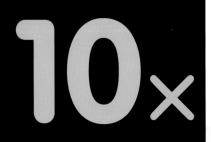

The ten times table

You don't need to memorize this times table. All you need to learn is the simple pattern in the numbers.

Just add zero

To make a number ten times larger, you add a zero to the end. This means that the units in the number become tens, and if there are any tens they become hundreds.

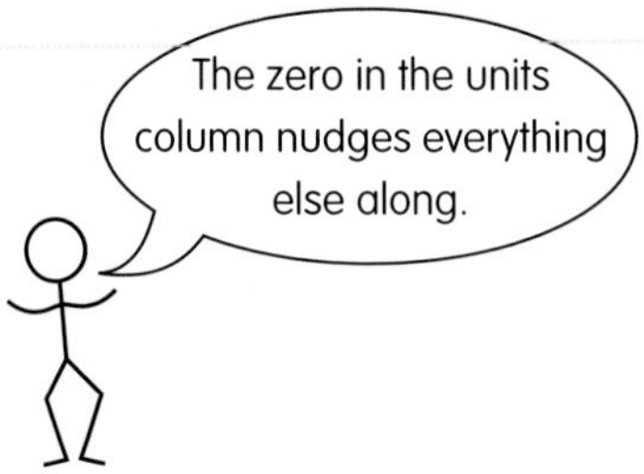
The zero in the units column nudges everything else along.

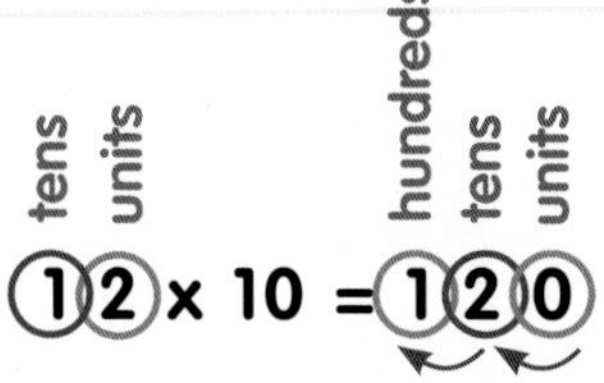
tens
units
hundreds
tens
units
12 × 10 = 120

There are **10** pencils in each group. How many altogether?

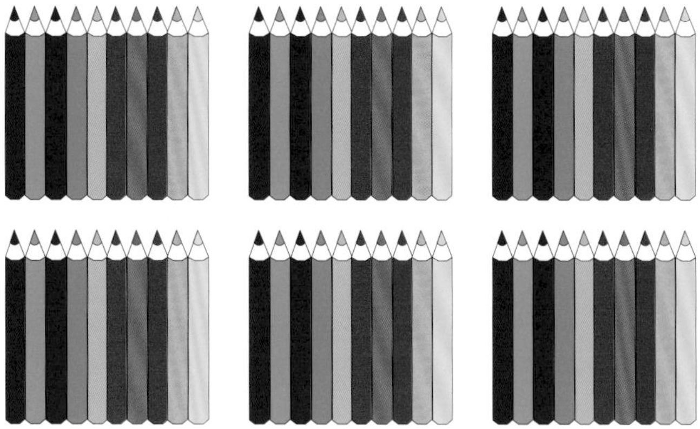

How many paperclips in these four groups of **10**?

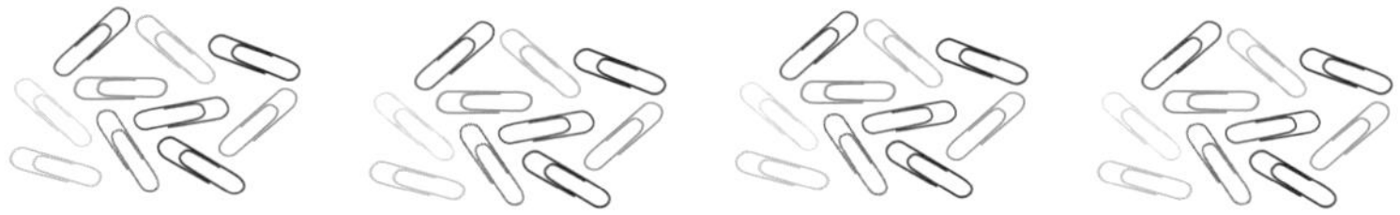

Now try multiplying these large numbers by **10**.

73 × 10 =
135 × 10 =
245 × 10 =

Can you work out what **451,236** multiplied by ten equals?

Answers: 60 pencils, 40 paperclips. 73 × 10 = 730. 135 × 10 = 1,350. 245 × 10 = 2,450. 451,236 × 10 = 4,512,360.

Let's go bowling

For each pin that you knock over, you score **10** points. Can you tell how many points you would score in each of these examples?

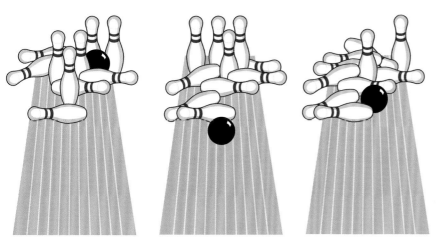

Hundreds and thousands

It's just as easy to multiply by **100** or **1,000**. To multiply a whole number by **100**, add two zeros to the end. To multiply a whole number by **1,000**, add three zeros. Make sure you add the same number of zeros to the answer as there are in the multiplier.

In this game, you score points for each alien invader that you disintegrate. Can you work out how many points have been scored in each example?

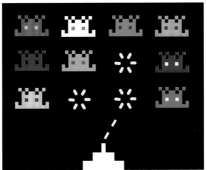

Each alien is worth 100 points.

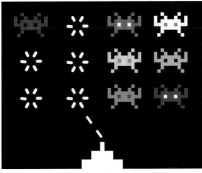

Each alien is worth 1,000 points.

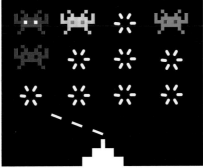

Each alien is worth 1,000 points.

The final countdown

It's easy to count down from **10** to **1**. But how quickly can you count down in tens?

100
90
80
70
60
50
40
30
20
10
Blast off!

49

10×

Give me ten!
multiplication game

This fast-paced two-player game will help you practise your times tables up to **10 × 10**.

TOP TIP
To multiply by numbers ending with a **0**, break the problem into two steps.

Imagine you want to multiply 50 by 6.
STEP 1 Multiply by the first part of the number, ignoring the zero.

$$5\cancel{0} \times 6 = 30$$

STEP 2 Now multiply by **10**.

$$30 \times \textcircled{10} = 300$$

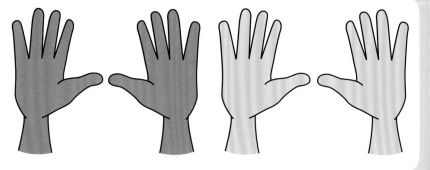

Each player must think of a number between **0** and **10** in their head. They both hold out their hands and call out, "Ready, steady, go!" then hold up that number of fingers.

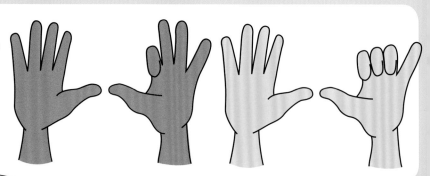

Now each player tries to work out the answer to the number of fingers held up on their hands, multiplied by the number of fingers held up on the other player's hands.

Whoever calls out the correct answer first wins a point. Keep playing until one player has won ten points.

Hey big spender
To multiply money by **10**, move the decimal point along to the right and add a zero at the end of the pence column. So for example:

£8.50 × 10 = £85.00

Make a times tables slider

This simple make-and-do project will turn times tables practice into a fun quiz game.

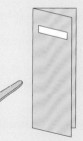

1) Take an A4-sized piece of coloured card, and fold it in two lengthwise. Cut a rectangular hole, about 1.5 cm (½ in) high and 5 cm (2 in) wide. If you like, you could decorate it with felt pens or stickers.

2) Now take a coloured piece of paper, and cut it into a strip 28 cm (11 in) long and 9 cm (3½ in) wide.

3) Write the questions and answers to a times table down the length of the strip of paper. The answer to each question should be written underneath it.

4) Now put the strip of paper inside the folded card, and move it about until the first question appears in the box. Say your answer out loud, then check it by pulling the strip upwards.

```
10 x 1 =
        10
10 x 2 =
        20
10 x 3 =
        30
10 x 4 =
        40
10 x 5 =
        50
10 x 6 =
        60
10 x 7 =
        70
10 x 8 =
        80
10 x 9 =
        90
10 x 10 =
        100
10 x 11 =
        110
10 x 12 =
        120
```

Make a different strip of paper for each times tables that you want to learn!

Picnic puzzler

If forks come in packs of **6**, and knives come in packs of **10**, how many packs of forks and how many packs of knives would you need to buy in order to have the same number of each?

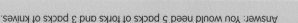

Answer: You would need 5 packs of forks and 3 packs of knives.

The four times table

If you already know the two times table well, you won't find it hard to learn the four times table. Remember to look for the patterns in the answers.

4×

Have you spotted the patterns?

Here's the 4 times table:

$1 \times 4 = 4$
$2 \times 4 = 8$
$3 \times 4 = 12$
$4 \times 4 = 16$
$5 \times 4 = 20$
$6 \times 4 = 24$
$7 \times 4 = 28$
$8 \times 4 = 32$
$9 \times 4 = 36$
$10 \times 4 = 40$
$11 \times 4 = 44$
$12 \times 4 = 48$

*Yes, look – all the answers in the four times tables are **EVEN** numbers.*

*There's another pattern too… As you read down the table the answers end **4, 8, 2, 6, 0**, over and over.*

Counting in fours

Many everyday objects come in groups of four. You can use the four times table to count them.

A car has 4 wheels

A chair has 4 legs

A dog has 4 legs

Count these in groups of four

How many wheels on **8** cars?

$8 \times 4 = 32$

How many legs on **4** chairs?

How many legs on **6** dogs?

Double, double – a lot less trouble!

The simplest way to multiply by **4** is to multiply by **2**, then multiply by **2** again.

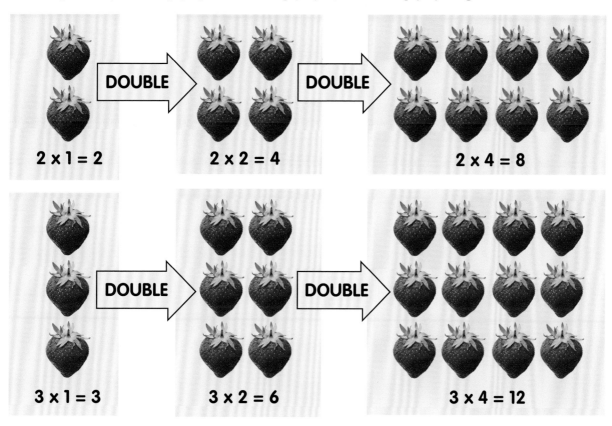

2 x 1 = 2   DOUBLE   2 x 2 = 4   DOUBLE   2 x 4 = 8

3 x 1 = 3   DOUBLE   3 x 2 = 6   DOUBLE   3 x 4 = 12

Four times table number grid

If you circle all the multiples of **4** in this grid, a pattern appears.

Can you tell which numbers will be circled after **48**?

(0)	1	2	3	(4)	5	6	7	(8)	9
10	11	(12)	13	14	15	(16)	17	18	19
(20)	21	22	23	(24)	25	26	27	(28)	29
30	31	(32)	33	34	35	(36)	37	38	39
(40)	41	42	43	(44)	45	46	47	(48)	49
50	51	52	53	54	55	56	57	58	59
60	61	62	63	64	65	66	67	68	69
70	71	72	73	74	75	76	77	78	79
80	81	82	83	84	85	86	87	88	89
90	91	92	93	94	95	96	97	98	99

Answer: the next circled numbers will be 52, 56, 60, 64, 68, 72, 76, 80, 84, 88, 92, 96.

How division works

You can use the times tables to divide. Dividing is like multiplying backwards, starting with the answer and ending with the question.

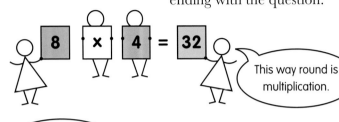

This way round is multiplication.

This way round is division.

Sharing between 4

Division is the same as sharing. **12 ÷ 4** means the same as **12** shared between **4.**

Imagine you're sharing **12** marbles between **4** friends. One way to do this is to count how many groups of **4** there are in **12**. But using the times tables is faster.

Dividing by 4 made easy

There's an simple shortcut when it comes to dividing by **4**.

16 cakes

REMEMBER, REMEMBER

One good way of memorizing a times table is to turn it into a rhyme. You could start like this:

**"Four times one is four,
I am going to the store.
Four times two is eight,
I just hope I'm not too late..."**

When you've made up your rhyme, keep repeating it out loud until you know it by heart.

First divide the number in half.
(This is the same as dividing by **2**.)

Then divide it in half again.

8 cakes

4 cakes

4 cakes

½

8 cakes

½

4 cakes

4 cakes

½

Calculator corner

44,444

Type these sums into a calculator, and write the answers down.

4 × 4 =
44 × 4 =
444 × 4 =
4444 × 4 =
44444 × 4 =

You can keep going if you like. Can you spot the surprising pattern in your answers?

TOP TIP

If you multiply two **even** numbers together, you get an **even** answer. Multiply two **odd** numbers together, and you get an **odd** answer. An **odd** number times an **even** number gives an **even** answer.

EVEN x EVEN = EVEN

ODD x ODD = ODD

ODD x EVEN = EVEN

Piggy banks

Divide the money in these piggy banks between **4** children. How much will they each receive?

36p

16p

48p

24p

16 ÷ 4 = 4

55

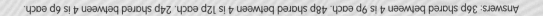

The eleven times table

Here's the 11 times table:

$1 \times 11 = 11$
$2 \times 11 = 22$
$3 \times 11 = 33$
$4 \times 11 = 44$
$5 \times 11 = 55$
$6 \times 11 = 66$
$7 \times 11 = 77$
$8 \times 11 = 88$
$9 \times 11 = 99$
$10 \times 11 = 110$
$11 \times 11 = 121$
$12 \times 11 = 132$

A helpful pattern

There is a simple way of multiplying single-digit numbers by **11**. Imagine you're multiplying **3** by **11**. Just think of the same number again, written next to it: **33**.

Another pattern

There is another pattern that will help you remember the three-digit answers for **10 × 11**, **11 × 11**, and **12 × 11**. It also works with other problems up to **18 × 11**.

The first and last digits in the answer to **10 × 11** are **1** and **0**.

Add the first and last numbers of the answer together to get the middle number: **1 + 0 = 1**.

$10 \times 11 = 1\,0$
$10 \times 11 = 110$

$12 \times 11 = 1\,2$
$12 \times 11 = 132$

The first and the last number in the answer to **12 × 11** are **1** and **2**.

Add the first and last numbers of the answer together to get the middle number: **1 + 2 = 3**.

Let's go fly a kite

Can you work out the answers to **10 × 11** to **18 × 11**? Follow the strings to see whether you're right.

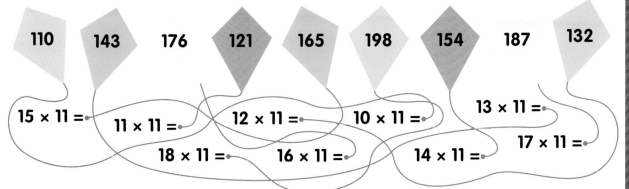

110	143	176	121	165	198	154	187	132

15 × 11 =

11 × 11 =

12 × 11 =

10 × 11 =

13 × 11 =

18 × 11 =

16 × 11 =

14 × 11 =

17 × 11 =

Football tournament

Footballers play in teams of **11** players.

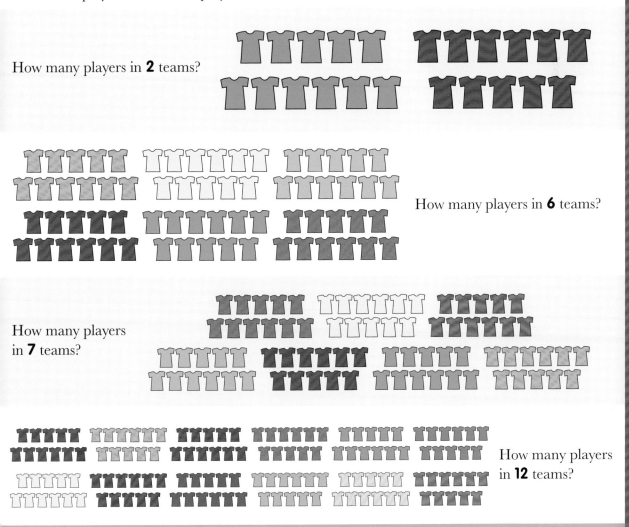

How many players in **2** teams?

How many players in **6** teams?

How many players in **7** teams?

How many players in **12** teams?

11×

Halftime drinks and snacks

You're buying the halftime refreshments for your local football team. There are **11** people on the team. How much will it cost you to buy enough of these for the whole team? Remember that **100p** is **£1**.

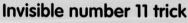

40p **70p** **50p** **40p** **80p**

Invisible number 11 trick

This magic trick will amaze your friends.

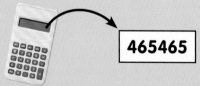

1. Dip a paintbrush in lemon juice and write the number **11** on a piece of paper. When the juice dries, the answer will be invisible. Show this blank paper to your audience.

2. Ask a volunteer to secretly think of a three-digit number.

465

3. Ask them to type the number into a calculator twice. If the number they thought of was **465**, they should type:

465465

4. First tell them to divide that number by lucky number **7**.

$$465{,}465 \div 7 = 66{,}495$$

5. Now tell them to divide it by the least lucky number: **13**.

$$66{,}495 \div 13 = 5{,}115$$

6. Finally, tell them to divide by the number they first thought of.

$$5{,}115 \div 465 = 11$$

7. Tell your audience that you are going to magically write the answer on the blank paper. Hold the paper near a hot lightbulb, and the number **11** will magically appear!

Ta-dah!

Wow!

11

Times tables pairs game

This two-player game will help you learn your times tables, at the same time as improving your memory.

1. First, you'll need to cut out **24** pieces of card. Choose **12** sums you find difficult, and write the questions on half of the cards, and the answers on the other half. Leave one side of each card blank.

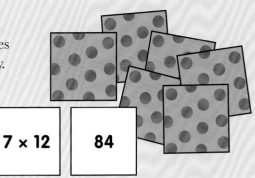

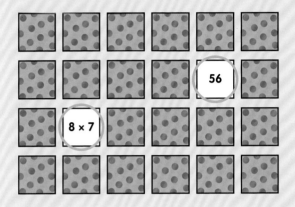

2. Shuffle the cards together, and spread them out face-down on the table in rows and columns, without looking at them.

3. Each player takes it in turns to turn over two cards. If the two cards they turn over are a matching pair showing a question and the correct answer, then they keep them. Otherwise they turn them face down again.

4. When there are no cards left, the winner is whoever has the most cards.

Watch the cards your opponent turns over, and try to remember them for your turn.

Prime time

The number **11** is a prime number. This means that it is only divisible by two natural numbers: **1** and itself. The number **1** is not a prime number. These are also prime numbers:

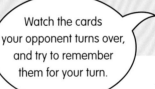

2   **3**   **5**   **7**

Can you work out which is the next prime number after **11**?

TOP TIP

If you ever have any problems multiplying large numbers by **11**, just remember: multiply by **10**, then add the original number.

Calculator corner

$11 \times 11 =$

Type these equations into your calculator, and an interesting pattern will appear.

$11 \times 11 =$
$111 \times 111 =$
$1{,}111 \times 1{,}111 =$
$1{,}1111 \times 1{,}1111 =$

Can you guess what the next number in the pattern will be?

$11{,}111 \times 11{,}111 = ?$

Answer: The next prime number is 13.

The three times table

There is no sneaky shortcut to learning the three times table – this one takes practice. But once you have mastered it, you will have learnt most of your times tables.

Counting in threes

Many objects with three parts start with "tri" – like triangles, tricycles and triplets. Can you count in threes?

A triangle has 3 sides   A tricycle has 3 wheels   A triplet has 3 notes

Count these in groups of three

How many wheels on **4** tricycles?

$4 \times 3 = 12$

How many sides on **7** triangles?

How many notes in **9** triplets?

Don't forget:
if you multiply **3** by an **odd** number, the answer will be an **odd** number.

But if you multiply **3** by an **even** number, the answer is an **even** number.

Answers: 21 sides, 27 triplets.

Carrot patch number line

You can also look at the **3** times table as a number line. Look at this rabbit – each time she jumps, she skips over two carrots and takes a nibble at the third. Try closing your eyes and saying out loud where she will land each time. Can you bounce right up to carrot number **36**?

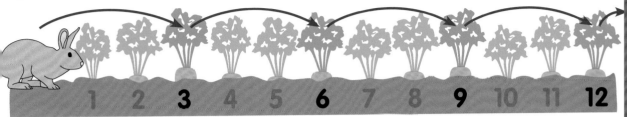

Rows and columns

How many balls are there of each colour? Find out by multiplying the rows and columns together. You could check your answers by counting the marbles.

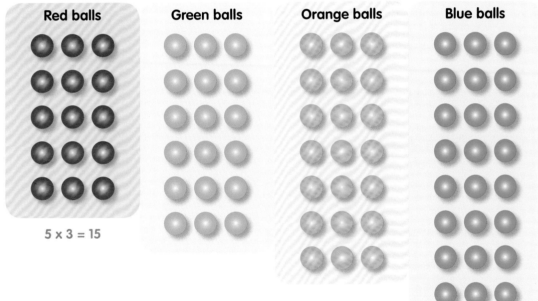

Red balls   **Green balls**   **Orange balls**   **Blue balls**

5 x 3 = 15

Times table farmyard

This is a noisy game for two or more players. Take it in turns to call out numbers, counting from **1**. Whenever someone reaches a multiple of three (an answer in the **3×** table), they have to make an animal noise.

1   2   Woof!   4   5   Baa!   7

61

Answers: 18 green balls, 21 orange balls, 24 blue balls.

3×

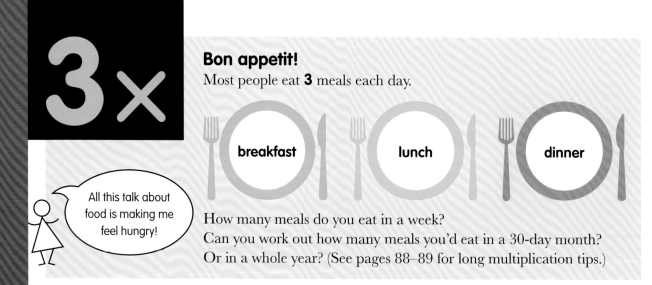

Bon appetit!
Most people eat **3** meals each day.

breakfast   lunch   dinner

All this talk about food is making me feel hungry!

How many meals do you eat in a week?
Can you work out how many meals you'd eat in a 30-day month?
Or in a whole year? (See pages 88–89 for long multiplication tips.)

Minesweeper

Can you find a safe route through this minefield? Start where it says "**go**" and move across, up, or down until you reach "**end**". But you must avoid the mines! Any square with a multiple of three has a mine under it.

GO	2	5	13	21	19	30	12	24	6
6	11	27	8	32	10	15	26	18	7
25	36	1	24	3	29	4	17	9	END
31	8	22	16	30	18	21	14	31	20

The solution is on page 96.

It all adds up

The first three answers in the **3×** table are **3**, **6**, and **9**. If you add together the digits of multiples of three, they add up to **3**, **6**, or **9**. You can use this as a way of checking your answers.

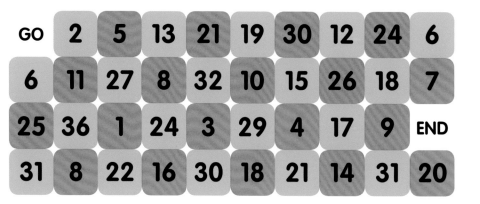

With 27 → 2 + 7 = 9

With 15 → 1 + 5 = 6

Answers: 21 meals in a week. 90 meals in a 30-day month. 1,095 meals in a non-leap year.

Magic number 3 card trick

Here's another magic trick you can perform for a friend. You'll need a deck of cards and a calculator.

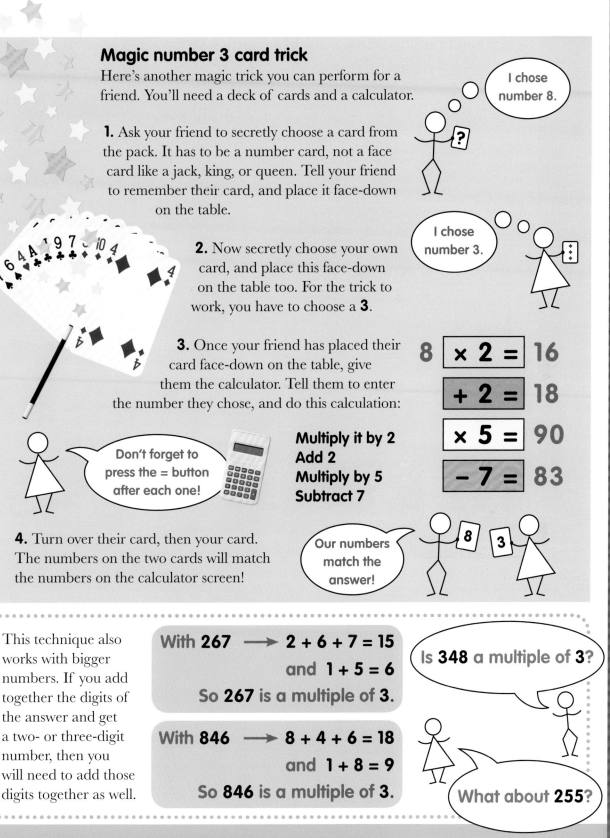

I chose number 8.

1. Ask your friend to secretly choose a card from the pack. It has to be a number card, not a face card like a jack, king, or queen. Tell your friend to remember their card, and place it face-down on the table.

I chose number 3.

2. Now secretly choose your own card, and place this face-down on the table too. For the trick to work, you have to choose a **3**.

3. Once your friend has placed their card face-down on the table, give them the calculator. Tell them to enter the number they chose, and do this calculation:

$$8 \quad \boxed{\times 2 =} \quad 16$$
$$\boxed{+ 2 =} \quad 18$$
$$\boxed{\times 5 =} \quad 90$$
$$\boxed{- 7 =} \quad 83$$

Don't forget to press the = button after each one!

Multiply it by 2
Add 2
Multiply by 5
Subtract 7

4. Turn over their card, then your card. The numbers on the two cards will match the numbers on the calculator screen!

Our numbers match the answer!

8 3

This technique also works with bigger numbers. If you add together the digits of the answer and get a two- or three-digit number, then you will need to add those digits together as well.

With **267** ⟶ 2 + 6 + 7 = 15
and 1 + 5 = 6
So **267** is a multiple of 3.

With **846** ⟶ 8 + 4 + 6 = 18
and 1 + 8 = 9
So **846** is a multiple of 3.

Is **348** a multiple of **3**?

What about **255**?

Answers: 348 and 255 are both multiples of 3.

How well do you know your time tables so far?

You can use this quiz to test yourself on the times tables you've learnt up to now. If you get stuck on some problems, make sure you review those tables later. The answers are on page 96.

Talking times tables

There are many ways of talking about the times tables. Can you answer each question?

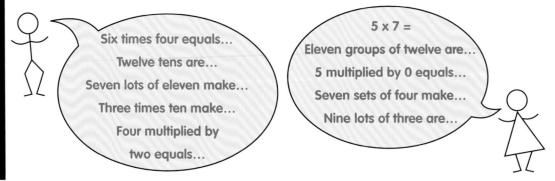

Six times four equals...

Twelve tens are...

Seven lots of eleven make...

Three times ten make...

Four multiplied by two equals...

5 x 7 =

Eleven groups of twelve are...

5 multiplied by 0 equals...

Seven sets of four make...

Nine lots of three are...

How many wheels?

11 bicycles   4 tricycles   5 trucks   12 cars   6 motorbikes

Big city buildings

How many windows are there on each of these buildings?

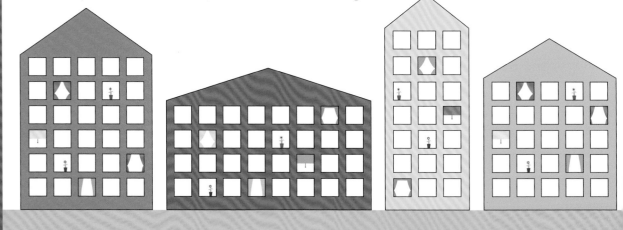

Fruit salad

You're making a fruit salad. How much will it cost you to buy…

25p 4 oranges

16p 6 bananas

31p 11 apples

82p 1 watermelon

63p 2 pineapples

Puzzle grid

Copy this grid onto a blank piece of paper. In each box, write the answer to the numbers along the top multiplied by the numbers along the side.

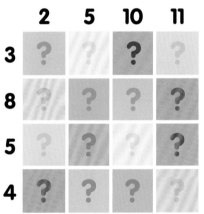

	2	5	10	11
3	?	?	?	?
8	?	?	?	?
5	?	?	?	?
4	?	?	?	?

Chilly aliens

These alien visitors to Earth are getting cold, so you've brought them some nice warm clothes. How many aliens will these go round?

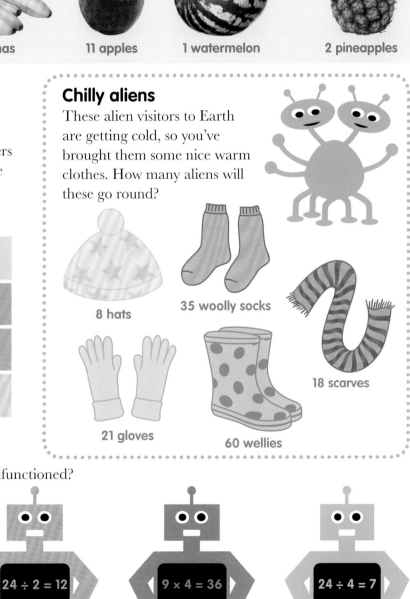

8 hats

35 woolly socks

18 scarves

21 gloves

60 wellies

Error! Error!

Which of these robots has malfunctioned?

9 × 11 = 99

24 ÷ 2 = 12

9 × 4 = 36

24 ÷ 4 = 7

65

The nine times table

The nine times table may look tricky, but it is one of the easiest tables to learn. There is a simple pattern hidden in the answers.

9×

Have you spotted the pattern?

Here's the 9 times table:

$1 \times 9 = 9$

$2 \times 9 = 18$

$3 \times 9 = 27$

$4 \times 9 = 36$

$5 \times 9 = 45$

$6 \times 9 = 54$

$7 \times 9 = 63$

$8 \times 9 = 72$

$9 \times 9 = 81$

$10 \times 9 = 90$

$11 \times 9 = 99$

$12 \times 9 = 108$

Look: the digits in the units column count down from **9** to **0**.

And between **2 × 9** and **10 × 9**, the tens column counts up from **1** to **9**.

Kitty multiplication

People sometimes say that cats have nine lives. How many lives do the cats below have between them?

1 cat = 9 lives

Count the lives for each group of cats

How many lives for **3** cats?

$3 \times 9 = 27.$

How many lives for **7** cats?

How many lives for **9** cats?

Nine times table number grid

0	1	2	3	4	5	6	7	8	9
10	11	12	13	14	15	16	17	18	19
20	21	22	23	24	25	26	27	28	29
30	31	32	33	34	35	36	37	38	39
40	41	42	43	44	45	46	47	48	49
50	51	52	53	54	55	56	57	58	59
60	61	62	63	64	65	66	67	68	69
70	71	72	73	74	75	76	77	78	79
80	81	82	83	84	85	86	87	88	89
90	91	92	93	94	95	96	97	98	99

Look what happens if you put the answers to the **9×** table in a grid.

The answers form a diagonal line!

The nine times table made easy

This is a rule that works for all the answers in the nine times table between **2 × 9** and **10 × 9**. The key is to do some quick mental subtraction.

STEP 1

The answers start with a digit one less than the number you are multiplying by **9**.

STEP 2

The second digit of the answer is equal to **9** minus the first digit.

2 - 1 = 1

2 × 9 = 18

2 × 9 = 18

9 - 1 = 8

3 - 1 = 2

3 × 9 = 27

3 × 9 = 27

9 - 2 = 7

Cats and kittens

If **9** cats each have **9** kittens, how many cats will there be in all?

TOP TIP

Here's an easy way of finding out whether a number is a multiple of nine. You can use this to check your answers. If a number is a multiple of nine, then the digits in the answer will add up to nine.

36 → 3 + 6 = **9**

This works with bigger numbers, too. If you add together the numbers in the answer and get a two-digit number, then you need to add those digits together too.

99 → 9 + 9 = **18**

1 + 8 = **9**

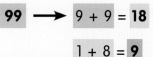

Can you work out whether **783** and **16,947** are multiples of nine?

A handy way of multiplying by nine

Here's a way to work out the **9×** table on your hands.

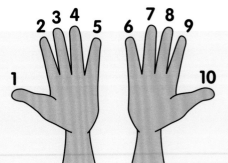

Hold your hands in front of you, palms upwards. Imagine that each of your fingers has a number written on it, from **1** to **10**.

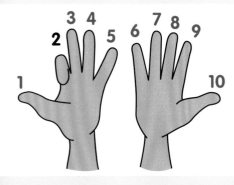

Say you want to multiply **9** by **2**. Start at the left and count along your fingers until you get to the second one. Fold that finger down.

How many fingers are to the left of the finger that's folded down? This is the first digit of your answer. How many fingers are to the right of the finger that's folded down? That's the second digit of your answer.

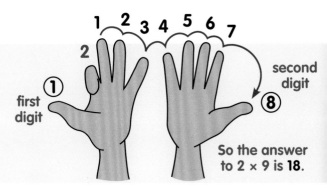

first digit

second digit

So the answer to 2 × 9 is **18**.

Now let's try some more.

What is 6 × 9?

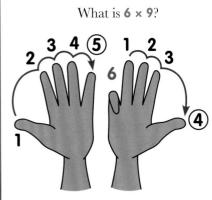

What is 4 × 9?

What is 9 × 9?

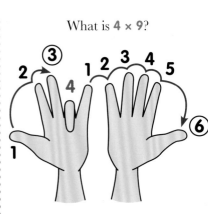

Answers: 6 × 9 = 54, 4 × 9 = 36, and 9 × 9 = 81.

Reversible answers

For each of the answers in the **9×** table, there is another answer with the digits swapped round. For example, **63** is a multiple of nine, but so is **36**, which is **63** reversed. Which multiplications are linked by their reversible answers? Follow the lines to find out. Can you say what the answers are?

2 × 9
(18)

3 × 9

4 × 9

5 × 9

7 × 9

6 × 9

8 × 9

9 × 9
(81)

Nine times brainteaser

Cut out ten pieces of card, and write these numbers on them:

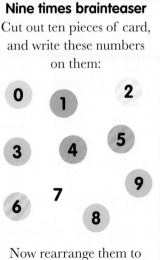

0 1 2 3 4 5 6 7 8 9

Now rearrange them to make **5** answers to equations in the **9** times table.

Incredible number nine magic trick

Write the number **9** on a piece of paper, slip it inside a balloon, and blow the balloon up.

1. Ask a volunteer to pick a three-digit number.

2. Jumble up the three digits in any way to make another number.

3. Take the smaller number away from the larger one.

4. Add the digits in the answer together. If the solution has more than one digit, keep adding the digits together until you have one number.

5. Now tell your audience that you are going to magically write that number on the piece of paper inside the balloon. Pop the balloon, and show them what is written inside. They will be amazed!

943

394

943 - 394 = 549

5 + 4 + 9 = 18

1 + 8 = 9

Ta-dah!

Answer to brainteaser: 90, 18 or 81, 27 or 72, 36 or 63, and 45 or 54.

69

The six times table

At this point, you have already learnt nine of the 13 times tables in this book. After you've mastered the six times table, there are only three more to go.

If you multiply **6** by an even number, they both end with the same digit.

Here's the 6 times table:

$1 \times 6 = 6$

$2 \times 6 = 12$

$3 \times 6 = 18$

$4 \times 6 = 24$

$5 \times 6 = 30$

$6 \times 6 = 36$

$7 \times 6 = 42$

$8 \times 6 = 48$

$9 \times 6 = 54$

$10 \times 6 = 60$

$11 \times 6 = 66$

$12 \times 6 = 72$

$8 \times 6 = 4\,8$

I've spotted something else, too! For the first four of those equations, the first digit is half of the second digit.

$2 \times 6 = 1\,2$

$6 \times 6 = 3\,6$

Counting in sixes

Many groceries come in groups of six. You can count them using the six times table.

A string of **6** sausages

A box of **6** eggs

Yoghurts come in packs of **6**

Count these in groups of six

If you have **2** strings of **6** sausages, how many sausages do you have?

If you have **8** boxes of **6** eggs, how many eggs do you have?

$2 \times 6 = 12$

How many yoghurts are here? Count the rows and columns.

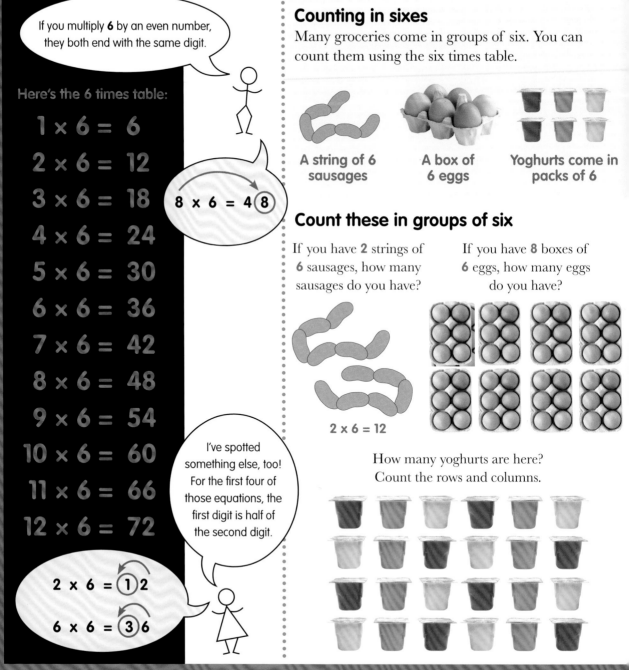

Taking the plunge

By multiplying the length of a rectangle by its width, you can find its area. Each of these swimming pools is **6** metres wide, but they each have a different length. Can you find their areas in metres squared?

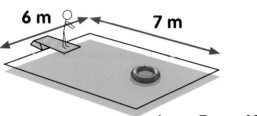

6 m × 7 m = 42 m²

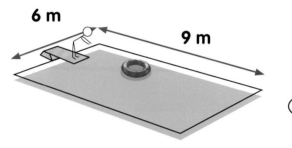

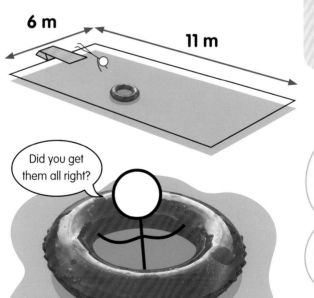

Did you get them all right?

Dotty dominoes game

This two-player game will help you practise your times tables up to **6×**. Place some dominoes face down on a table, and mix them up. Take it in turns to turn one over. When it's your turn, you must multiply the two sides of the domino together and say the answer out loud. If you get it right, you can keep the domino. If you're wrong, turn it back over.

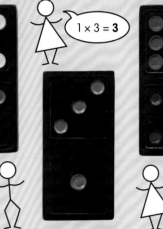

1 × 3 = **3**

5 × 3 = **15**

6 × 4 = **24**

Whoever has the most dominoes at the end of the game wins.

TOP TIP

If you get stuck with the **6x** table, remember that it's the same as the **3x** table **doubled**.

7 × 6 is the same as

7 × 3, **twice**.

Or you can take the **5x** table as a starting point, and add **1** more set.

4 × 6 is the same as

4 × 5, + another **4**.

6×

How bee-wildering

These multiplications from the **6×** table have got all mixed up. Can you work out what they should say?

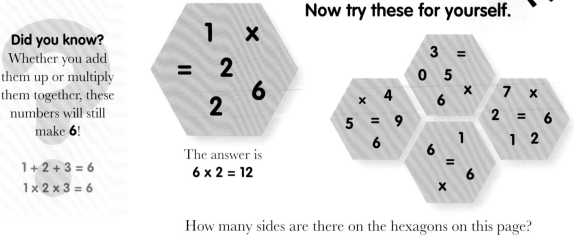

1 × = 2 2 6

The answer is
6 × 2 = 12

Now try these for yourself.

3 = 0 5 × 4 6 × 7 × 5 = 9 6 2 = 6 6 1 1 2 = 6 ×

How many sides are there on the hexagons on this page?

Sandcastles

You have built **6** sandcastles, and want to divide the decorations you have equally between them. How many of each decoration will you put on each castle?

18 flags   **24 starfish**   **48 shells**   **66 pebbles**

18 ÷ 6 = 3
so there will be
3 flags
on each sandcastle

Answers: How bee-wildering: 6 × 5 = 30, 6 × 9 = 54, 6 × 12 = 72, 6 × 1 = 6, 3 flags, 4 starfish, 8 shells, 11 pebbles, 30 sides.

The finger calculator

Here's a brilliant way of multiplying together numbers between **6** and **9**, using only your fingers. All you need to know to do this are the **1** to **4** times tables. This technique will help you to check your answers for some of the trickiest multiplication problems.

Imagine you want to multiply together **8** and **6**.
8 × 6 = ?

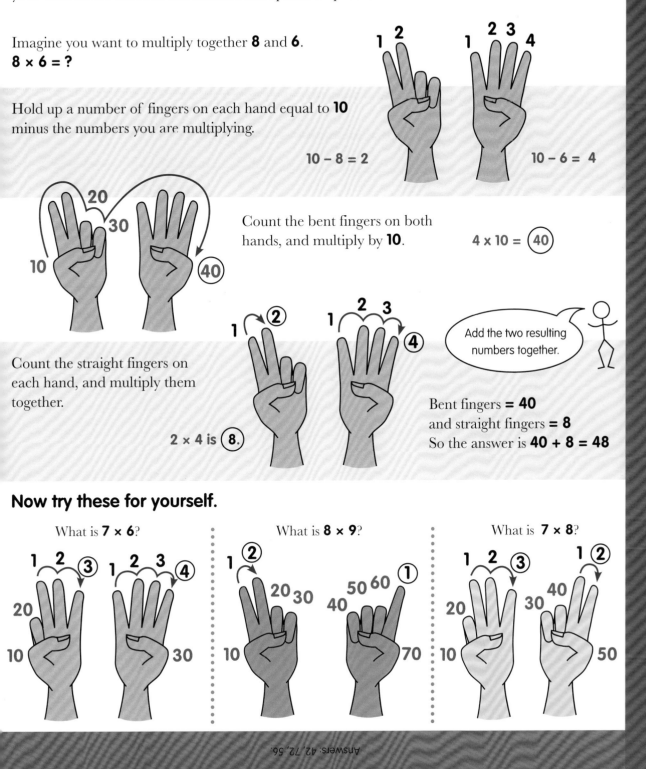

Hold up a number of fingers on each hand equal to **10** minus the numbers you are multiplying.

10 − 8 = 2   10 − 6 = 4

Count the bent fingers on both hands, and multiply by **10**.

4 × 10 = 40

Count the straight fingers on each hand, and multiply them together.

2 × 4 is 8.

Add the two resulting numbers together.

Bent fingers **= 40**
and straight fingers **= 8**
So the answer is **40 + 8 = 48**

Now try these for yourself.

What is **7 × 6**?   What is **8 × 9**?   What is **7 × 8**?

The seven times table

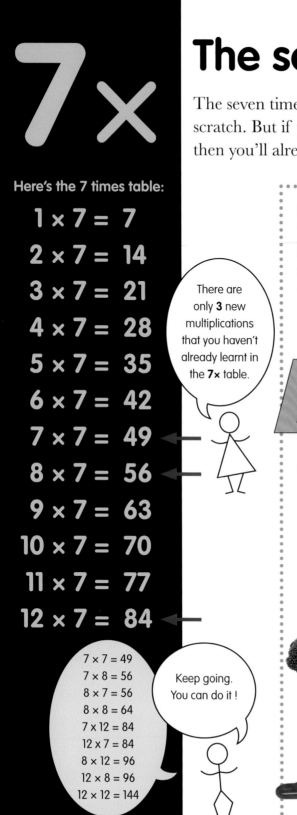

The seven times table is one of the hardest to learn from scratch. But if you've learnt all the previous times tables, then you'll already know most of the sevens.

7×

Here's the 7 times table:

1 × 7 = 7
2 × 7 = 14
3 × 7 = 21
4 × 7 = 28
5 × 7 = 35
6 × 7 = 42
7 × 7 = 49
8 × 7 = 56
9 × 7 = 63
10 × 7 = 70
11 × 7 = 77
12 × 7 = 84

There are only **3** new multiplications that you haven't already learnt in the **7×** table.

7 × 7 = 49
7 × 8 = 56
8 × 7 = 56
8 × 8 = 64
7 × 12 = 84
12 × 7 = 84
8 × 12 = 96
12 × 8 = 96
12 × 12 = 144

Keep going. You can do it !

Every day of the week

The seven times table is useful for counting how many times things happen in a week, or over several weeks.

7 Sunday

1 Monday **2** Tuesday **3** Wednesday **4** Thursday **5** Friday **6** Saturday

Count these in groups of seven

If you eat **5** fruits or vegetables a day, how many pieces is that a week?

If you wash your hand **6** times each day, how many times will you wash them in a week?

5 × 7 = 35

If you brush your teeth twice a day, how many times is this a week?

Fairytale division

The **7** dwarves have dug up these treasures in their mine, and want to share them out equally. Using division, can you work out how many each dwarf should receive?

They find **63** emeralds. How many does each dwarf get?

They find **35** gold nuggets. How many does each dwarf get?

They find **84** rubies. How many does each dwarf get?

63 ÷ 7 = 9

Count-around

Here's a trick that will help you remember **4 × 3** and **7 × 8**. If you follow the arrows around this sum, the numbers read " **1, 2, 3, 4.**"

4 × 3 = 12

The same thing works for 7 × 8! Just think: "5, 6, 7, 8."

8 × 7 = 5 6

Odd ones out

Which of these are not multiples of **7**?

63

36

14

81

70

56

21

There are seven stripes in a rainbow.

How many stripes are there in 7 rainbows?

What about 70 rainbows?

Or **700** rainbows?

Answers: 49 stripes, 490 stripes, 4,900 stripes.

75

Patterns in the seven times table

Look at the pattern in this grid. These are the **first digits** for each answer (up to **9 × 7**) in the **7×** table.

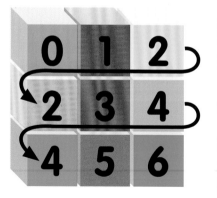

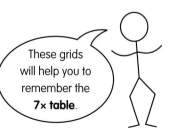

These grids will help you to remember the **7× table**.

The number at the end of each is also the first number of the next line.

0, 1, 2 … 2, 3, 4 … 4, 5, 6

Now try this one.

There's another special pattern in this grid. Start at the top right and read down, and you'll see the numbers **1** to **9**.

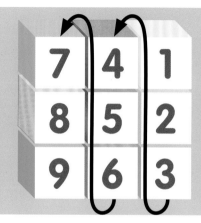

This grid gives the **final digits** for each answer (up to **9 × 7**) in the **7×** table.

What happens if we put our two grids together?

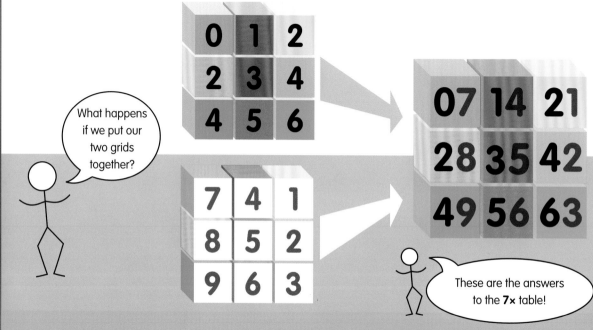

These are the answers to the **7×** table!

Gift boxes

You can find out the amount of space inside each of these boxes by multiplying together their length, height, and width. The answer will be in centimetres cubed (cm³).

$3 \text{ cm} \times 2 \text{ cm} \times 7 \text{ cm} =$

First, multiply **3** by **2**.

$3 \times 2 = 6$

Multiply the answer by **7**.

$6 \times 7 = 42\text{cm}^3$

How many cm³?

How about this one?

And this one?

REMEMBER, REMEMBER

You can practise your times tables by making flashcards. Cut out **12** pieces of card about **10 cm** by **5 cm** (**4 in** by **2 in**). Write the questions from a times table on one side, and the answers on the other. Then go through the problems on the cards, saying the answers as quickly as possible before turning the cards over.

Write the questions on the front.

7 x 3   21

7 x 7   49

7 x 12   84

When you have practised for a while, separate the cards into two piles: hard questions, and easy questions. Then practise the ones that you find hardest.

Write the answers on the back.

Answers: 28 cm³, 350 cm³, 140 cm³.

8×

The eight times table

There are several helpful patterns in the eight times table that will help you to learn it quickly.

Have you spotted the patterns?

Here's the 8 times table:

1 × 8 = 8
2 × 8 = 16
3 × 8 = 24
4 × 8 = 32
5 × 8 = 40
6 × 8 = 48
7 × 8 = 56
8 × 8 = 64
9 × 8 = 72
10 × 8 = 80
11 × 8 = 88
12 × 8 = 96

All the answers end in **even numbers.**
The units of the answers count down in **twos**.
8, 6, 4, 2, 0. 8, 6, 4, 2, 0. Get the picture?

Counting in eights

Octopus arms and spider legs come in groups of eight. You can use the **8×** table to count them.

8 arms

8 legs

Count the arms and legs

How many arms on **4** octopuses?

4 x 8 = 32

How many legs on **7** spiders?

How many arms on **9** octopuses?

Answers: 56 spider legs, 72 octopus arms.

Game of chess

The **2** players in a game of chess each have **2** rows of **8** pieces. How many pieces are there altogether?

A chessboard has **8 × 8** squares. How many squares are there in total?

Odd ones out

Which of these numbers are not multiples of **8**?

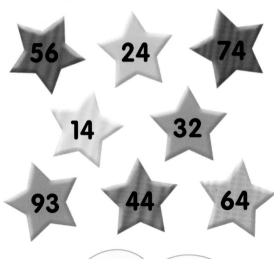

56 24 74

14 32

93 44 64

TOP TIP

If you can't remember a multiplication in the eight times table, remember that the answers in the eight times table are double the answers in the four times table.

6 × 4 = 24

And **24 × 2 = 48**

So **6 × 8 = 48**

4 Times tables tennis

You have to think fast in this two-player game.

1. First decide which times table you're going to practise (for example the **8×** table). Then decide which player is going to "serve", and which one will "return".

2. The server calls out numbers between one and twelve. The returning player must call back the answer to that number multiplied by **8** (or whichever times table you've chosen).

3. As soon as the returning player hesitates or gets an answer wrong, the players swap round, and the returning player starts to serve.

5! 40! 3! 24!

Answers: 32 chess pieces, and 64 squares. 74, 14, 93, and 44 are not multiples of 8.

8×

Division with remainders

When you can't divide a number equally, some will be left over. This leftover portion is called a remainder. To divide **19** by **8**, count down from **19** until you find a number that is in the **8** times table.

16 is the nearest multiple of **8**.

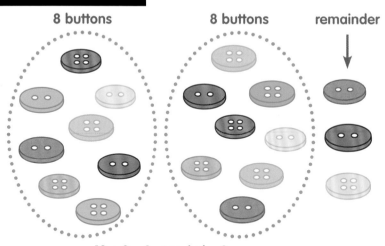

8 buttons   8 buttons   remainder

19 ÷ 8 = 2, remainder 3

19, 18, 17, 16...

How many times does **8** go into **16**?

The answer is 2.

How many are left over as a remainder?

19 − 16 = 3

What is the answer to **25 ÷ 8**?

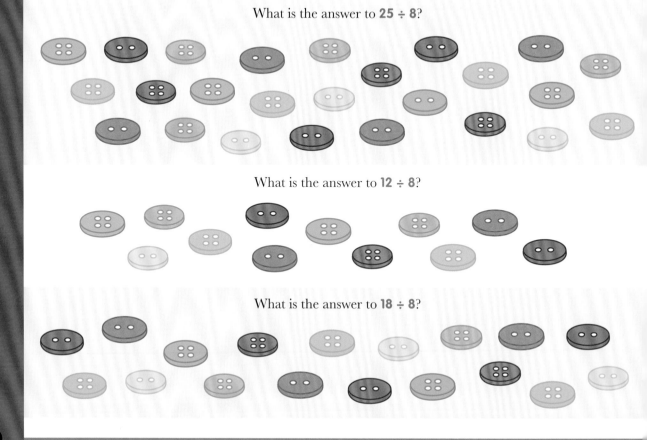

What is the answer to **12 ÷ 8**?

What is the answer to **18 ÷ 8**?

Monkey puzzle

There are **8** monkeys in the zoo's monkey enclosure, and the keeper has some crates full of different kinds of fruit. She wants to give each monkey the same amount of fruit, and this will mean that some fruit is left in each crate.

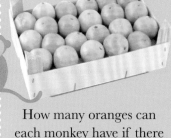

How many apples can each monkey have if there are **26** apples in all? How many will be left over?

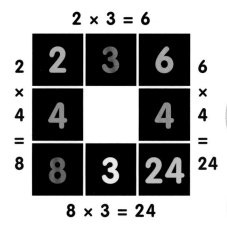

How many oranges can each monkey have if there are **76** oranges altogether? How many will be left over?

How many bananas can each monkey have if there are **41** bananas altogether? How many will be left over?

Puzzle squares

Copy these puzzle squares onto a blank piece of paper. Can you work out which numbers are missing from each square? The first two numbers in each row or column must multiply together to give the last number in that row or column.

2 × 3 = 6

	2 × 3 = 6	
2	**2**	**3**	**6**	**6**
×	**×**		**×**	**×**
4	**4**		**4**	**4**
=	**=**		**=**	**=**
8	**8**	**3**	**24**	**24**

8 × 3 = 24

TOP TIP

Dividing by **8** can be tricky. It's much easier to halve a number. If you ever get stuck, instead of dividing by **8**, try halving, halving and halving again.

24	24
÷8	÷2
↓	↓
3	12
	÷2
	↓
	6
	÷2
	↓
	3

It works the other way round, too: you can multiply by **8** by doubling three times.

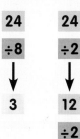

Fill in the missing squares. (Solution on page 96.)

2		4
8	16	

2		8
	4	48

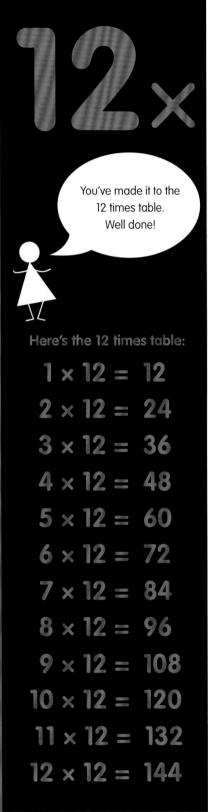

The twelve times table

This is the last times table to learn – or do you know it already? We've already covered **143** of the **144** equations from the **1** to **12** times tables, so now there should only be one equation you don't know.

You've made it to the 12 times table. Well done!

Here's the 12 times table:

$$1 \times 12 = 12$$
$$2 \times 12 = 24$$
$$3 \times 12 = 36$$
$$4 \times 12 = 48$$
$$5 \times 12 = 60$$
$$6 \times 12 = 72$$
$$7 \times 12 = 84$$
$$8 \times 12 = 96$$
$$9 \times 12 = 108$$
$$10 \times 12 = 120$$
$$11 \times 12 = 132$$
$$12 \times 12 = 144$$

The 12× table – fast

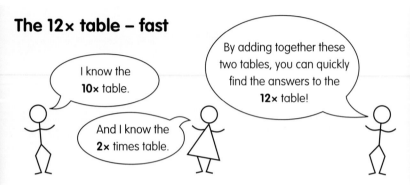

I know the **10×** table.

And I know the **2×** times table.

By adding together these two tables, you can quickly find the answers to the **12×** table!

How many candles on these **5** birthday cakes?

How many hearts are on these **3** wedding cakes?

How many coloured sweets on these **4** chocolate cakes?

Answers: 60 candles, 36 hearts, 48 coloured sweets.

Record-breaking dice game

Now that you have learnt the times tables up to **12 × 12**, you can play a simple game to revise them. Roll two dice, and make a mental note of the number you scored. Roll the dice again, and multiply the first number by the second. Did you get it right? Check your answer. See how many you can get right in a row – then try to beat your own record!

Quick-thinking dice game

Here's another dice game you can play with a friend. Roll two dice, then roll them again. You must multiply the first number you got by the second. Whoever calls out the right answer first scores a point. Keep playing until one player wins by scoring **10** points.

REMEMBER, REMEMBER

These are the hardest equations in the **12×** table:

11 × 12 = 132   **12 × 12 = 144**

Write each of these on a flash card. Then fix the cards to each side of your bedroom door, using sticky tack. Before you can open the door you must give the password – which is the answer on the back of the card.

Tied together

Which of these equations have the same answers? Untangle the strings to find out if you're right.

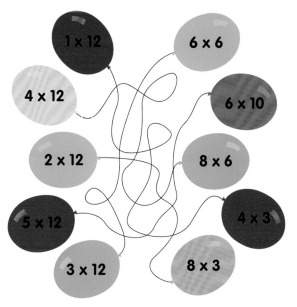

1 × 12   6 × 6   4 × 12   6 × 10   2 × 12   8 × 6   5 × 12   4 × 3   3 × 12   8 × 3

Whack-a-mole

If you bop a mole on the head, you score **12** times the number on its card. How many points is each mole worth? See if you can answer them all in less than **30** seconds.

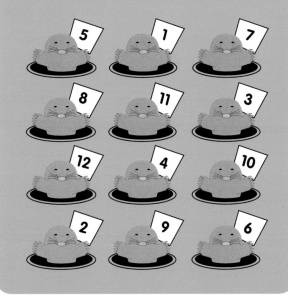

5   1   7
8   11   3
12   4   10
2   9   6

12×

Odd ones out

Which of these numbers are not multiples of **12**?

108   60   70   144

54   142   24

Number cards are worth **2–10** points.
Jacks count as **11** points.
Queens count as **12** points.
Kings count as **12** points.

The ace is
worth **1** point.

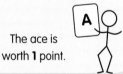

Times tables snap

This is a game for two players. You'll need a pack
of cards with the jokers taken out.

1) Shuffle the pack and deal
the cards face down between
the players. They should
each have a pile of **26** cards
in front of them.

2) Both players flip over the card on the top of their
deck. Whoever calls out the correct product first wins
both cards. (The product is the answer to those two
cards multiplied together.) They put the cards they've
won in a separate winnings pile.

3) If one of the players calls out the wrong
answer, the other player wins both cards.

4) If both players call out the correct answer
at the same time, that round is a draw, and
the players must turn over more cards until
there is a clear winner. All the cards turned
over then go to the winner of that round.

6 × 4 = 24
I win!

And the winner is...
The player with the most cards in their
winnings pile at the end of the game has won.

84

Times tables bingo

You can play this game with two or more friends – the more people, the better. One person has to be the caller, and the other people are the players.

1) First, each player needs to draw a grid of **25** squares on a piece of paper, like this (see right).

2) Then they write a number in each square. They can choose any numbers from the list below. The caller and other players mustn't look at what they are writing.

Some numbers on your card are more likely to be called than others. Can you work out why?

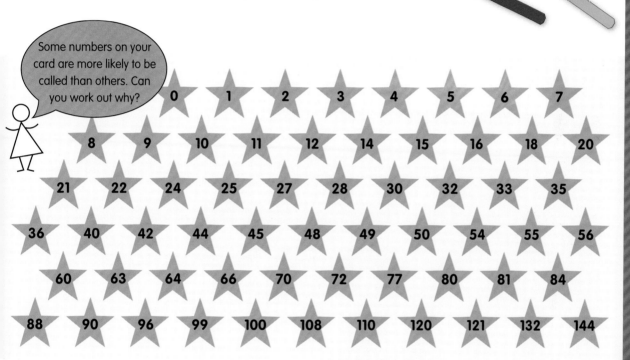

0 1 2 3 4 5 6 7
8 9 10 11 12 14 15 16 18 20
21 22 24 25 27 28 30 32 33 35
36 40 42 44 45 48 49 50 54 55 56
60 63 64 66 70 72 77 80 81 84
88 90 96 99 100 108 110 120 121 132 144

3) When the players have written down all their numbers, the caller starts to shout out problems from the **0** to **12** times tables. If a player has the answer to one of the equations on their sheet, they should cross it out.

And the winner is...
The first person to cross out all the answers on their sheet should shout out "BINGO!"

Bingo!

Count on me
Did you know it's possible to count from **1** to **12** on one hand? Touch the tip of your thumb to each of the joints of your fingers in turn.

How well do you know your times tables?

You can test yourself on all the tables up to **12×** with the puzzles here. Make a note of any problems you find especially difficult – that way you can practise them later. The answers are on page 96.

How many legs?

| 12 spiders | 7 elephants | 5 ladybirds | 9 ducks | 11 snakes |

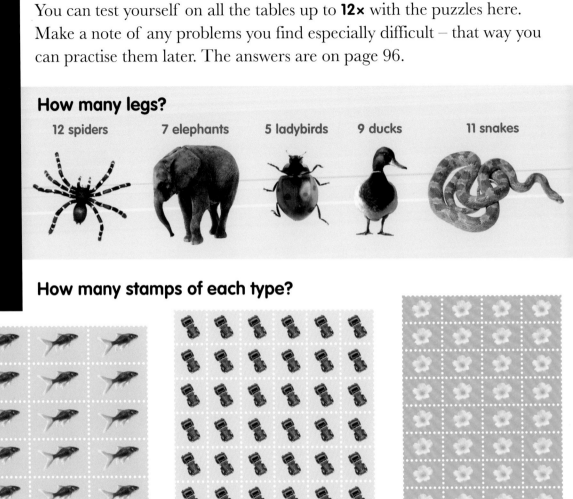

How many stamps of each type?

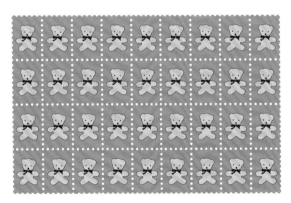

Dominoes

Multiply the two halves of these dominoes together.

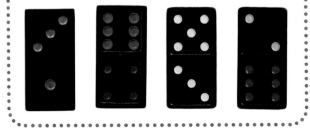

At the toy shop

You have **£4.68** in your piggy bank. How many of each of these could you afford? How much would be left as a remainder?

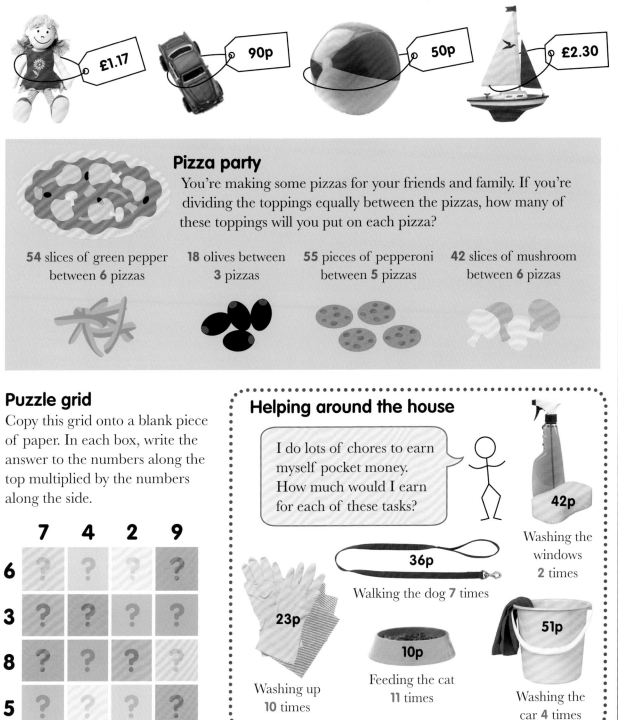

£1.17

90p

50p

£2.30

Pizza party

You're making some pizzas for your friends and family. If you're dividing the toppings equally between the pizzas, how many of these toppings will you put on each pizza?

54 slices of green pepper between **6** pizzas

18 olives between **3** pizzas

55 pieces of pepperoni between **5** pizzas

42 slices of mushroom between **6** pizzas

Puzzle grid

Copy this grid onto a blank piece of paper. In each box, write the answer to the numbers along the top multiplied by the numbers along the side.

	7	4	2	9
6	?	?	?	?
3	?	?	?	?
8	?	?	?	?
5	?	?	?	?

Helping around the house

I do lots of chores to earn myself pocket money. How much would I earn for each of these tasks?

42p
Washing the windows **2** times

36p
Walking the dog **7** times

23p
Washing up **10** times

10p
Feeding the cat **11** times

51p
Washing the car **4** times

Long multiplication

Don't you touch that calculator!

Multiplying a large number by a single-digit number

This isn't as hard as it looks, but you will need to know your times tables up to **10 × 10** pretty well before you try this.

Write the large number above the small one.

H T U
786
× 2

Multiply the single-digit number on the bottom by the units, then tens, then hundreds of the number at the top.

786   Multiply by units
× 2
12 → 6 × 2 = 12

786   Multiply by tens
× 2
12
160 → 80 × 2 = 160

786   Multiply by hundreds
× 2
12
160
1400 → 700 × 2 = 1,400

12
160
+ 1400   Finally, add up the answers to those three multiplications.

1572   So: 786 × 2 = 1,572

Now let's try the fast way

A quicker way of doing this is to write the answer to each multiplication on the same line, going from right to left. If you get an answer of ten or more when you're multiplying the units, tens, or hundreds, you "carry" the first digit of that answer, adding it to the column to the left.

285
× 3   5 × 3 = 15
5
1   Carry 1 to tens column.

285   8 × 3 = 24
× 3
55   24 + 1 = 25
2 1   Carry 2 to hundreds column.

285   2 × 3 = 6
× 3
855   6 + 2 = 8
2 1

Now have a go

385
× 2

723
× 4

210
× 3

974
× 8

Multiplying two large numbers together

If you are multiplying together two numbers that have more than one digit, things get a little trickier. Keep practising and you'll soon pick it up.

Time to get your thinking cap on.

First concentrate on the unit digit at the bottom, and multiply it by each number on the top row in turn.

H T U
824
× 36

Ignore this 3 at first.
Multiply 6 by 4, then 2, then 8.

824	824	824
× 36 6 × 4 = 24	× 36 6 × 2 = 12	× 36 6 × 8 = 48
4	44 12 + 2 = 14	4944 48 + 1 = 49
2 carry the 2	1 2 carry the 1	1 2

Now look at the tens digit at the bottom, and multiply it by the units, tens and hundreds digits in the top row. But first you need to add a zero, because you're multiplying by numbers in the tens column.

824	824	824
× 36 3 × 4 = 12	× 36 3 × 2 = 6	× 36 3 × 8 = 24
4944	4944 6 + 1 = 7	4944
20 Add a zero	720	24720
1 carry the 1	1	1

824
× 36
————
4944
+ 24720
————
29664

Finally, add together the two rows of numbers.

The answer is 29,664.

I got the right answer!

I think I'm getting the hang of it...

Now have a go

285
× 23

628
× 71

457
× 18

767
× 58

526
× 99

614
× 63

Window-frame multiplication

Here's another way of multiplying large numbers together.
Some people find this easier than standard long multiplication.

Say, for example, you want to multiply 45 by 6.

A) The number **45** has **2** digits, so draw **2** rectangular boxes side by side.

B) Draw a diagonal line across each box, from the bottom left-hand corner to the top right-hand corner.

C) Write the numbers you want to multiply along the top and right-hand side of the boxes.

The number **6** is only one digit long, so only one row of boxes is needed.

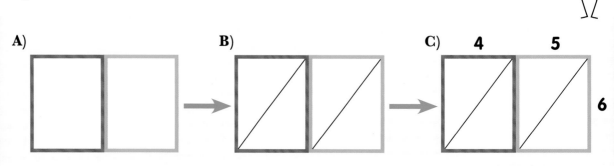

D) Multiply the digits along the top and side, starting from the right. **5 × 6 = 30**, so write **3** and **0** on either side of the diagonal line.

E) Now do the multiplication in the next box along. **4 × 6 = 24**, so write **2** and **4**.

F) Look at the numbers in each diagonal column. These give you the answer to **45 × 6**. If there are two numbers in a diagonal column, add them together.

Add together the numbers in the purple column.

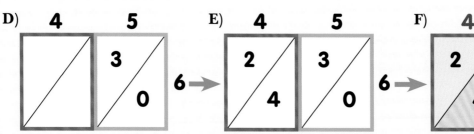
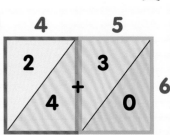

45 × 6 = 270

Window-frame multiplication works for larger numbers, too. Read the answers down the left-hand side and across the bottom of the boxes.

What is **24 × 32**?

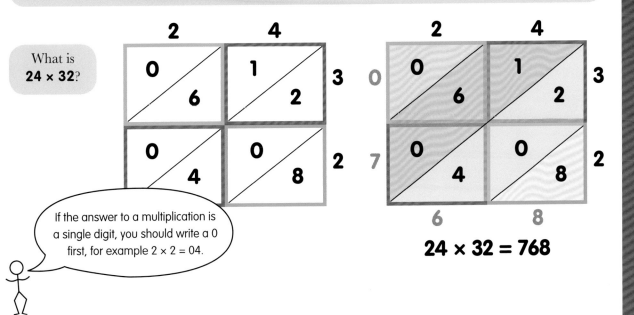

If the answer to a multiplication is a single digit, you should write a 0 first, for example 2 × 2 = 04.

24 × 32 = 768

If a diagonal column adds up to a two-digit answer, you should carry the first digit, adding it to the number on the left.

What is **34 × 28**?

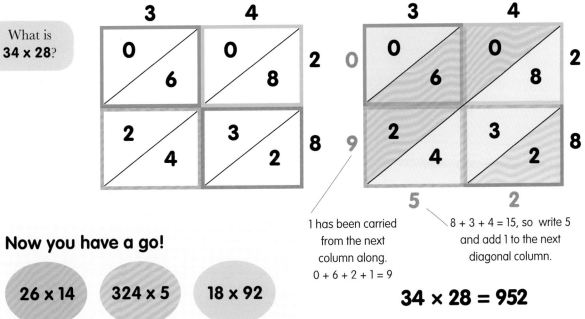

1 has been carried from the next column along.
0 + 6 + 2 + 1 = 9

8 + 3 + 4 = 15, so write 5 and add 1 to the next diagonal column.

34 × 28 = 952

Now you have a go!

26 × 14   324 × 5   18 × 92

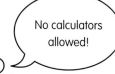

No calculators allowed!

Long division

Division is something that we use all the time, and you won't always have a calculator at hand. So it's worth learning how to divide large numbers with just a pen and paper.

Short division

Short division means dividing a large number by a one-digit number.

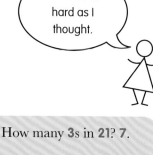

This isn't as hard as I thought.

Write **651 ÷ 3** like this:

$$3\overline{)651}$$

1) Divide each digit of the large number by the one-digit number, from left to right.

2) If you get a remainder, put this in front of the next digit along.

How many **3**s in **6**? **2**.

$$3\overline{)6\,5\,1}^{\,2}$$

(You're really dividing 600 by 3.)

How many **3**s in **5**? **1**, with a remainder of **2**.

$$3\overline{)6\,5^2 1}^{\,2\,1}$$ ← Remainder

(You're actually dividing 50 by 3.)

How many **3**s in **21**? **7**.

$$3\overline{)6\,5^2 1}^{\,2\,1\,7}$$

Now have a go!

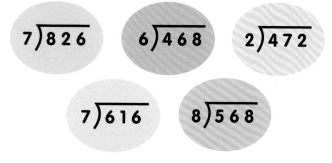

$$7\overline{)826}$$ $$6\overline{)468}$$ $$2\overline{)472}$$

$$7\overline{)616}$$ $$8\overline{)568}$$

TOP TIP

Another way to work out division problems is to use multiplication and guesswork. This is called trial and error.

Imagine you're trying to divide 108 by 3.

$40 \times 3 = 120$. This is too big.

$30 \times 3 = 90$. This is too small.

$35 \times 3 = 105$. Nearly there...

$36 \times 3 = 108$.

What happens if you try to divide a number by zero?

It doesn't matter how many zeros you add together – they will never add up to a whole number. You could go on adding

$0 + 0 + 0 + 0$ forever.

This is why any number divided by zero equals **infinity**.

Long division

This is a bit trickier. Make sure your brain is in gear before you tackle these problems!

Write **4081 ÷ 13** like this:

$$13\overline{)4081}$$

Now have a go!

How many times does **13** go into **4**? None, so move along one digit.

$$13\overline{)4081}$$
Work from left to right.

$$36\overline{)845}$$

How many times does **13** go into **40**? **3** times – so put a **3** above the **0**.

3 times **13** is only **39**. So we subtract **39** from **40** to get the remainder: **1**.

$$
\begin{array}{r}
3 \\
13\overline{)4081} \\
-39 \\
\hline
1
\end{array}
$$

13 goes into 40 3 times.

3 × 13 = 39

Remainder

$$24\overline{)5361}$$

Now you need to deal with the **8**. Move it down alongside the **1**, to make **18**.

How many times does **13** go into **18**? **1** time – so put a **1** above the **8**.

1 times **13** is **13**. Subtract this from **18** to find the remainder – **5**.

$$
\begin{array}{r}
31 \\
13\overline{)4081} \\
-39 \\
\hline
18 \\
-13 \\
\hline
5
\end{array}
$$

13 goes into 18 once.

1 × 13 = 13

Remainder

$$13\overline{)823}$$

Move the **1** down.

How many times does **13** go into **51**? **3** times, with a remainder of **12**.

There are no more digits to carry down, so we're finished. Phew.

$$
\begin{array}{r}
313 \\
13\overline{)4081} \\
39 \\
\hline
18 \\
-13 \\
\hline
51 \\
39 \\
\hline
12
\end{array}
$$

13 goes into 51 3 times.

3 × 13 = 39

Final remainder

> The answer is **313**, remainder **12**.

$$23\overline{)4810}$$

$$18\overline{)417}$$

> I think I need a break now!

93

Times tables grid

Use this grid to check your answers.

To find the product of a multiplication, simply trace your fingers along the row and column matching each of the numbers you want to multiply together, until you find the point where they meet.

×	1	2	3	4	5	6	7	8	9	10	11	12
1	1	2	3	4	5	6	7	8	9	10	11	12
2	2	4	6	8	10	12	14	16	18	20	22	24
3	3	6	9	12	15	18	21	24	27	30	33	36
4	4	8	12	16	20	24	28	32	36	40	44	48
5	5	10	15	20	25	30	35	40	45	50	55	60
6	6	12	18	24	30	36	42	48	54	60	66	72
7	7	14	21	28	35	42	49	56	63	70	77	84
8	8	16	24	32	40	48	56	64	72	80	88	96
9	9	18	27	36	45	54	63	72	81	90	99	108
10	10	20	30	40	50	60	70	80	90	100	110	120
11	11	22	33	44	55	66	77	88	99	110	121	132
12	12	24	36	48	60	72	84	96	108	120	132	144

This line is called the leading diagonal. The answers on either side of the line are mirror images of each other.

Glossary

Here are some important times tables words and phrases.

Don't get stuck – look it up!

Area how we measure the size of a surface. Area is measured in square units, for example, square metres.

Carry move a digit from one column to another in an addition or multiplication equation.

Difference what is left after one number is taken away from another.

Digits the symbols that make up numbers. For example, **25** is made up of the digits **2** and **5**.

Dividend a number that is divided by another in a division equation.

Division splitting a number into equal parts. One example of division is sharing between people. Division is the opposite of multiplication.

Divisible can be divided into a whole number, without a remainder. For example, **8** is divisible by **4**, because **8 ÷ 4 = 2**, and **2** is a whole number.

Divisor a number by which another number is divided.

In the equation **20 ÷ 5 = 4**, **5** is the divisor.

Even number a whole number that can be divided by **2** without a remainder. Even numbers end with the digits **0, 2, 4, 6,** or **8**.

Factor whole numbers that can be multiplied together to make another number. For example **3** and **6** are factors of **18**.

Multiple a number that can be divided by another number without a remainder. For example, **54** is a multiple of **6** because **9 × 6 = 54**.

Multiplication adding the same number over and over again.

Multiplier a number that is multiplied by another number.

Odd number a whole number that cannot be divided by **2** without a remainder. Odd numbers end with the digits **1, 3, 5, 7,** or **9**.

Prime number a number that only has two factors: **1** and itself.

Product the result of a multiplication. In the equation **3 × 5 = 15**, **15** is the product.

Remainder when you have divided a whole number into smaller whole numbers, what is left over is called the remainder.

Units the last digit in a whole number. For example, in **513** the unit is **3**.

Volume a measurement of how much a three-dimensional shape could contain, measured in units cubed, for example, metres cubed.

Whole number a number that does not end in a decimal or fraction.

Learn as many as you can!

WELL DONE!

95

Answers

Times tables quiz
Pages 64–65

6	15	30	33	Puzzle grid
16	40	80	88	
10	25	50	55	
8	20	40	44	

Talking times tables
$6 \times 4 = 24$ $5 \times 7 = 35$
$12 \times 10 = 120$ $11 \times 12 = 132$
$7 \times 11 = 77$ $5 \times 0 = 0$
$3 \times 10 = 30$ $7 \times 4 = 28$
$4 \times 2 = 8$ $9 \times 3 = 27$

How many wheels?
22 wheels on 11 bicycles
12 wheels on 4 tricycles
20 wheels on 5 trucks
48 wheels on 12 cars
12 wheels on 6 motorbikes

Big city buildings
30 windows in the purple building, red building – 32, yellow building – 21, green building – 25

Fruit salad
£1.00 for 4 oranges
96p for 6 bananas
£3.41 for 11 apples
82p for 1 watermelon
£1.26 for 2 pineapples

Chilly aliens
There are enough hats for 4 aliens.
7 aliens can wear gloves.
9 aliens can wear scarves.
7 aliens can wear woolly socks.
12 aliens can wear wellies.

Error! Error!
The green robot has malfunctioned.

He should have said: $28 \div 4 = 7$.

Times tables quiz
Pages 86–87

42	24	12	54	Puzzle grid
21	12	6	27	
56	32	16	72	
35	20	10	45	

How many legs?
12 spiders have 96 legs.
7 elephants have 28 legs.
5 ladybirds have 30 legs.
9 ducks have 18 legs.
11 snakes have 0 legs.

How many stamps of each type?
15 fish stamps
36 car stamps
32 flower stamps
36 teddy bear stamps

Dominoes
$3 \times 1 = 3$ $5 \times 3 = 15$
$6 \times 4 = 24$ $2 \times 6 = 12$

At the toy shop
4 dolls (no remainder)
5 toy cars (18p remainder)
9 beach balls (18p remainder)
2 toy boats (8p remainder)

Pizza party
9 pieces of pepper, 6 olives, 11 pieces of pepperoni, 7 pieces of mushroom

Helping around the house
£2.30 for washing up
£2.52 for walking the dog
£1.10 for feeding the cat
£2.04 for washing the car
84p for washing the windows

8x table squares
Page 81

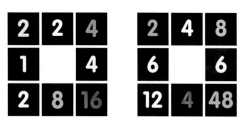

3x table minefield
Page 62

TiMES TaBLES

TIME CHALLENGES

Consultant Sean McArdle

Note for parents
Find a kitchen timer and allow your child 10 minutes for each activity in this chapter. Practising little and often will support their learning and keep maths fun!

Contents

Time filler:
In these boxes are some extra challenges to extend your skills. These can be stand-alone activities that you can do in 10 minutes. Why not ask someone to time you?

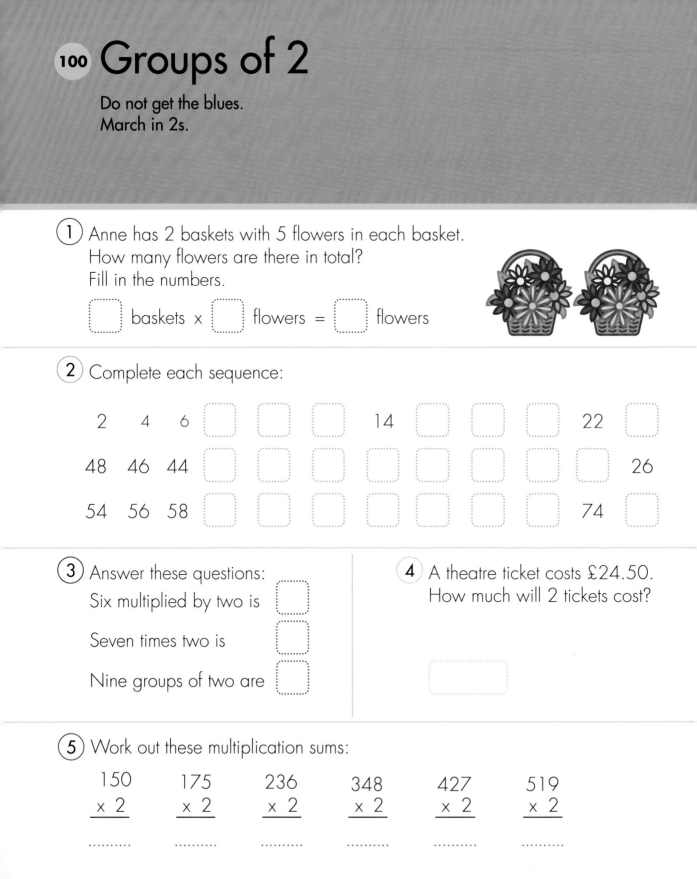

Do not get the blues.
March in 2s.

① Anne has 2 baskets with 5 flowers in each basket.
How many flowers are there in total?
Fill in the numbers.

☐ baskets x ☐ flowers = ☐ flowers

② Complete each sequence:

2   4   6   ☐ ☐ ☐   14   ☐ ☐ ☐   22   ☐

48   46   44   ☐ ☐ ☐ ☐ ☐ ☐ ☐   26

54   56   58   ☐ ☐ ☐ ☐ ☐ ☐   74   ☐

③ Answer these questions:

Six multiplied by two is ☐

Seven times two is ☐

Nine groups of two are ☐

④ A theatre ticket costs £24.50.
How much will 2 tickets cost?

☐

⑤ Work out these multiplication sums:

150	175	236	348	427	519
x 2	x 2	x 2	x 2	x 2	x 2
....	....	....	....	....	....

Time filler:
Can you recite the 2x table backwards?
Time yourself to see how quick you can be.

101

(6) Divide each number by 2:

76 ▢   142 ▢   178 ▢

(7) Work out these division sums:

▢
2)126

▢
2)240

▢
2)352

▢
2)684

▢
2)792

(8) Fazir and Tira shared £7.80 equally between them.
How much money did each child receive?

▢

(9) There were 284 bees in 2 hives. If there was an equal number
in each hive, how many bees were there in 1 hive?

▢ bees

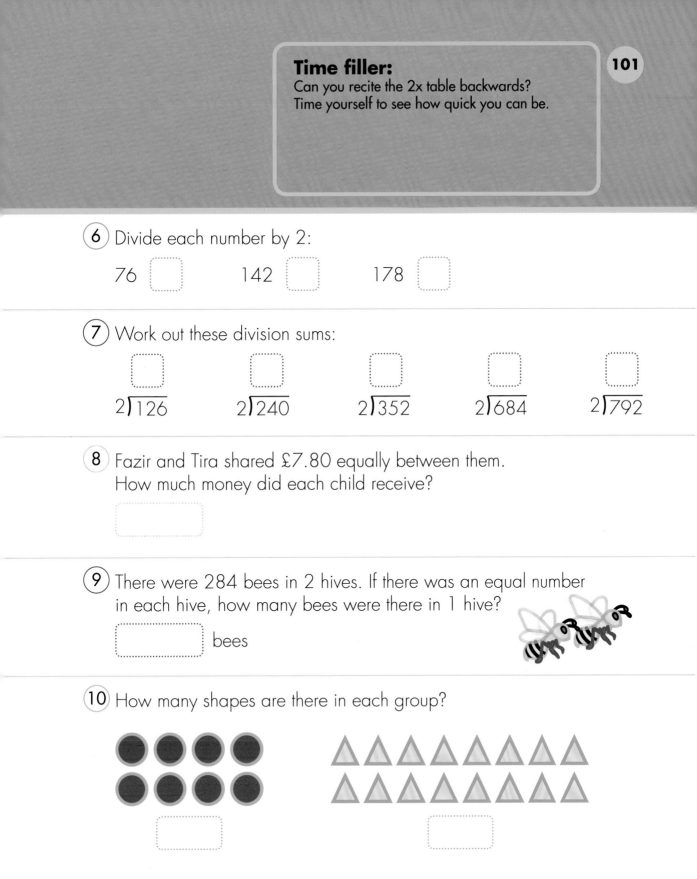

(10) How many shapes are there in each group?

▢   ▢

Pairs and doubles

Forget your troubles,
Forget your cares.
Practise doubles and pairs!

1 Double each number:

25 ⬚   42 ⬚   70 ⬚   127 ⬚

2 How many socks are there in 36 pairs?

⬚ socks

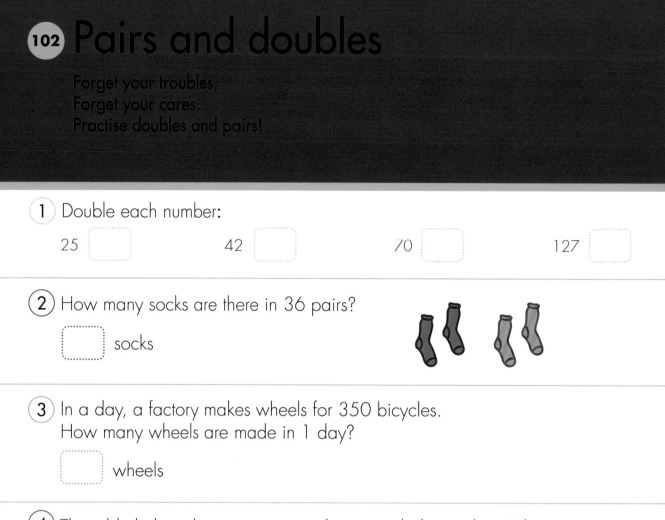

3 In a day, a factory makes wheels for 350 bicycles.
How many wheels are made in 1 day?

⬚ wheels

4 The table below shows some ingredients needed to make 12 biscuits.
Calculate how much of each you will need to make 24 biscuits.
Hint: Double each amount.

Ingredients	Quantity for 12 biscuits	Quantity for 24 biscuits
Flour	350g (12oz)	
Eggs	2	
Butter	225g (8oz)	
Caster sugar	175g (6oz)	
Dark chocolate	350g (12oz)	
Light brown sugar	175g (6oz)	

Time filler:
I am thinking of a number between 1 and 10.
I double it, double again and double yet again.
My answer is 24. What number did I start
with? Ask friends to do the same exercise
with a different starting number. Can you
work out what their starting number was?

103

5) How many wings do 275 crane flies have altogether?
Note: A crane fly has two wings.

[ ] wings

6) The chart shows the number of bunches
of flowers sold in a store in one week.
Write the total for each day.

✿ = 2 bunches

Day	Number of bunches sold	Total
Monday	10 x ✿	
Tuesday	8 x ✿	
Wednesday	12 x ✿	
Thursday	9 x ✿	
Friday	20 x ✿	
Saturday	14 x ✿	
Sunday	5 x ✿	

7) Ryan cycles 56 km, but Jake
cycles twice as far. How far
does Jake cycle?

[ ]

8) Mum has spent £84.00 on
presents for Jayden, but Dad
has spent double that amount.
How much money has
Dad spent?

[ ]

104 Groups of 10

Count up in 10s.
Again and again.

(1) Tiya had 7 parcels. Each parcel weighed 10kg.
How much did the parcels weigh altogether?

(2) Complete these sequences:

10   20   30

150 140 130

270 280 290

(3) Answer these questions:

Ten eights are

Ten times ten is

Nine multiplied by ten is

(4) Zina saved 35 10-pence coins.
How much money did Zina
have altogether?

(5) Work out these multiplication sums:

436	845	152	1 689	791	287
×10	×10	×10	×10	×10	×10

Time filler:
Think of a 2-digit number. Multiply the
number by 10, multiply the same number
by 20 and then the same number again by
30. Do you notice a pattern? To multiply your
number by 40, try multiplying by 10 then
multiplying the answer by 4.

105

(6) Divide each number by 10:

10 ⬚   40 ⬚   80 ⬚   120 ⬚   150 ⬚

(7) Work out these division sums:

$10\overline{)420}$ $10\overline{)367}$ $10\overline{)780}$ $10\overline{)842}$ $10\overline{)990}$

(8) How many leaves are there in each group? **Hint:** Multiply the number
of rows by the number of columns.

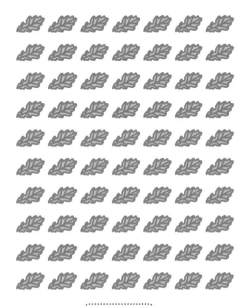

Multiplying by 100 and 1000

Carefully count the zeros
To be a math superhero!

(1) Multiply each number by 100:

4 [＿＿＿]   47 [＿＿＿]   470 [＿＿＿]   4 070 [＿＿＿]

(2) A box contains 100 T-shirts. How many T-shirts are there in 64 boxes?

[＿＿＿] T-shirts

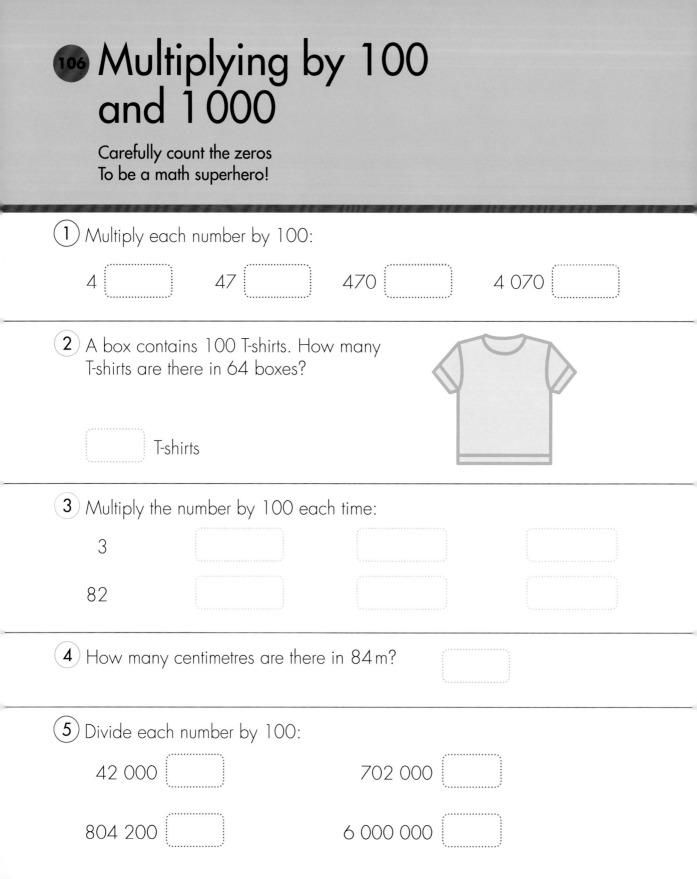

(3) Multiply the number by 100 each time:

3   [＿＿＿]   [＿＿＿]   [＿＿＿]

82   [＿＿＿]   [＿＿＿]   [＿＿＿]

(4) How many centimetres are there in 84 m?   [＿＿＿]

(5) Divide each number by 100:

42 000 [＿＿＿]   702 000 [＿＿＿]

804 200 [＿＿＿]   6 000 000 [＿＿＿]

Time filler:
What is 30% of £1 500?
Remember: 30% is the same as $\frac{30}{100}$.
So divide £1 500 by 100 and multiply by 30.
It is sale time. Calculate 20% of £80, 40% of £200, and 60% of £45.50.

6) Multiply each number by 1 000:

7 [ ]   82 [ ]   146 [ ]   150 [ ]

7) How many grams are there in 7.2 kg?

[ ]

8) How many pence are there in £35?

[ ]

9) A plane flies at a height of 10 668 m.
What is this height in kilometres?

[ ]

10) A colony of army ants has 700 000 ants. As the ants cross a river,
20% of the colony dies. How many ants make it across?

[ ] ants

Groups of 3

Count in groups of 3.
It is as easy as can be.

① A jar holds 8 biscuits. How many biscuits are there in 3 jars?

[   ] biscuits

② Complete each sequence:

0   3   6   [ ] [ ] [ ] [ ] [ ] [ ] [ ]

36   33   30   [ ] [ ] [ ] [ ] [ ] [ ] [ ]

36   39   42   [ ] [ ] [ ] [ ] [ ] [ ] [ ]

③ Answer these questions:

Three fives are [ ]

Three multiplied by seven is [ ]

Three times nine is [ ]

④ Neo bought 6 oranges at 30p each. What was the total cost of the 6 oranges?

[   ]

⑤ Work out these multiplication sums:

16	33	55	79	145	229
x 3	x 3	x 3	x 3	x 3	x 3
....	....	....	....	....	....

Time filler:
Say the 3x table to a rap beat. Singing the times tables helps to learn them. Try saying them to your own musical beats.

109

6 Divide each number by 3:

6 ☐   15 ☐   24 ☐   36 ☐   45 ☐

7 How long will it take Anita to save 42 p if she saves 3 p every week?   ☐ weeks

8 Work out these division sums:

☐   ☐   ☐   ☐   ☐

3)60   3)90   3)72   3)99   3)183

9 Pablo was paid £3 for each car that he washed. He earned £39 in one week. How many cars did Pablo wash that week?

☐ cars

10 How many shapes are there in each group?

☆ ☆ ☆ ☆ ☆ ☆ ☆   ☺ ☺ ☺ ☺ ☺ ☺ ☺ ☺ ☺
☆ ☆ ☆ ☆ ☆ ☆ ☆   ☺ ☺ ☺ ☺ ☺ ☺ ☺ ☺ ☺
☆ ☆ ☆ ☆ ☆ ☆ ☆   ☺ ☺ ☺ ☺ ☺ ☺ ☺ ☺ ☺

☐   ☐

Triple the fun!
Multiply x3 to get these done.

(1) How many wheels are there
on 15 tricycles?

[........] wheels

(2) How many sides do
55 triangles have?

[........] sides

(3) About 170 triplets are born in the United Kingdom each year.
How many babies is this?

[........] babies

(4) Thirty-nine trimarans race in a
competition. How many hulls
are there altogether? **Note:**
A trimaran is a boat with 3 hulls.

[........] hulls

(5) Fifty-four children are split into
groups of 3. How many groups
of children are there?

[........] groups

(6) A magnifying glass makes bugs look triple their size. Below are the original
sizes of the bugs. What size is each of the bugs when it is magnified?

Worm: 6.5 cm   [........]

Centipede: 5.25 cm   [........]

Ladybird: 1.75 cm   [........]

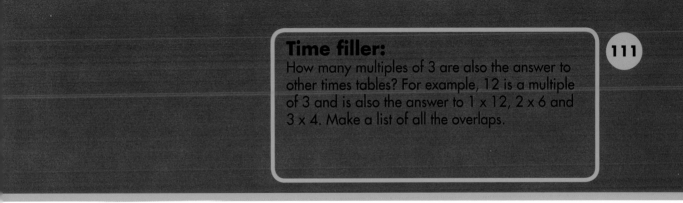

Time filler:
How many multiples of 3 are also the answer to other times tables? For example, 12 is a multiple of 3 and is also the answer to 1 x 12, 2 x 6 and 3 x 4. Make a list of all the overlaps.

111

(7) Packets of biscuits are sold in boxes of 3 packets. This chart shows how many boxes are sold from a store in a week. Calculate the number of packets sold that week.

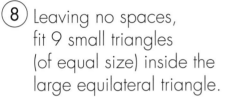 = 3 packets

Day	Number of boxes	Total
Monday		
Tuesday		
Wednesday		
Thursday		
Friday		
Saturday		
Sunday		

(8) Leaving no spaces, fit 9 small triangles (of equal size) inside the large equilateral triangle.

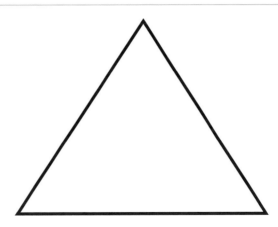

Are you ready for more?
Here is counting in sets of 4.

1) Share 28 sweets equally among 4 children.
How many sweets will each child get?

| | sweets

2) Complete each sequence:

0 4 8 [ ] [ ] [ ] [ ] [ ] [ ] [ ]

48 44 40 [ ] [ ] [ ] [ ] [ ] [ ] [ ]

52 56 60 [ ] [ ] [ ] [ ] [ ] [ ] [ ]

3) Answer these questions:

Nine times four is [ ]

Seven groups of four are [ ]

Four fives are [ ]

4) Dad took Devan, Jesse, and Owen to the fair. The roller coaster ride cost £1.50 for each person. How much did Dad have to pay for all of them to go on the ride?

[ ]

5) Work out these multiplication sums:

23	17	25	115	200	214
x 4	x 4	x 4	x 4	x 4	x 4
....	....	....	....	....	....

Time filler:
Another way to work out the answer to 4 times a number is to multiply the number by 2 and then its answer by 2 again. Choose some numbers between 1 and 20 and give this a go.

(6) Divide each number by 4:

0 ▢   4 ▢   16 ▢   36 ▢   48 ▢

(7) Jeff buys a pack of 4 pencils. The pack costs £1.68.   ▭
How much does 1 pencil cost?

(8) Work out these division sums:

▢ 4⟌56   ▢ 4⟌96   ▢ 4⟌100   ▢ 4⟌128   ▢ 4⟌284

(9) A box contains 24 chocolates. They are laid out in 4 equal rows.
How many chocolates are there in each row?

▢ chocolates

(10) How many shapes are there in each group?

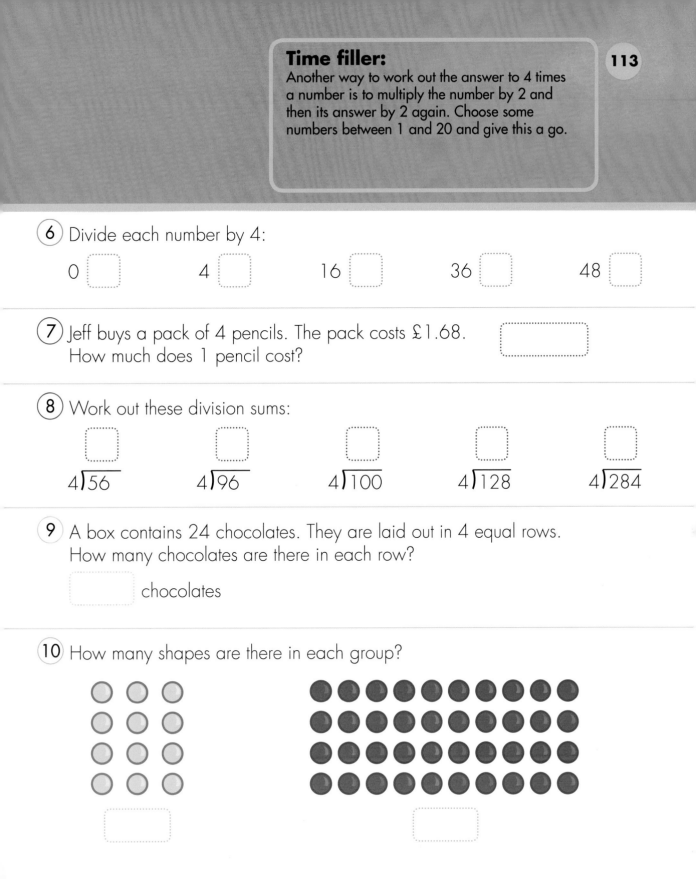

▭   ▭

Shapes

Count the angles and the sides,
Then read the question and multiply.

(1) How many sides do these shapes have in total?

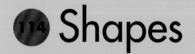

(2) How many triangles have 27 angles in total?

_____ triangles

(3) The inside angles of an equilateral triangle add up to 180°. What is the value of each angle?

_____

(4) Each side of a regular hexagon is 7 cm. What is the perimeter of the hexagon?

_____

(5) What is the area of a rectangle with a length of 11 cm and a breadth of 4 cm?

_____

(6) Each angle of a square is 90°. What is the total of the 4 angles?

_____

Time filler:
What times tables will help you solve these problems: What is the perimeter of a regular pentagon with 4-cm-long sides? What is the area of a square with 6-cm-long sides? What is the volume of a cube with 3-cm-long sides?

115

(7) A cuboid has 8 vertices. How many cuboids will have a total of 80 vertices?

[   ] cuboids

(8) How many faces do these shapes have?

7 triangular prisms [   ]

9 cuboids [   ]

12 cylinders [   ]

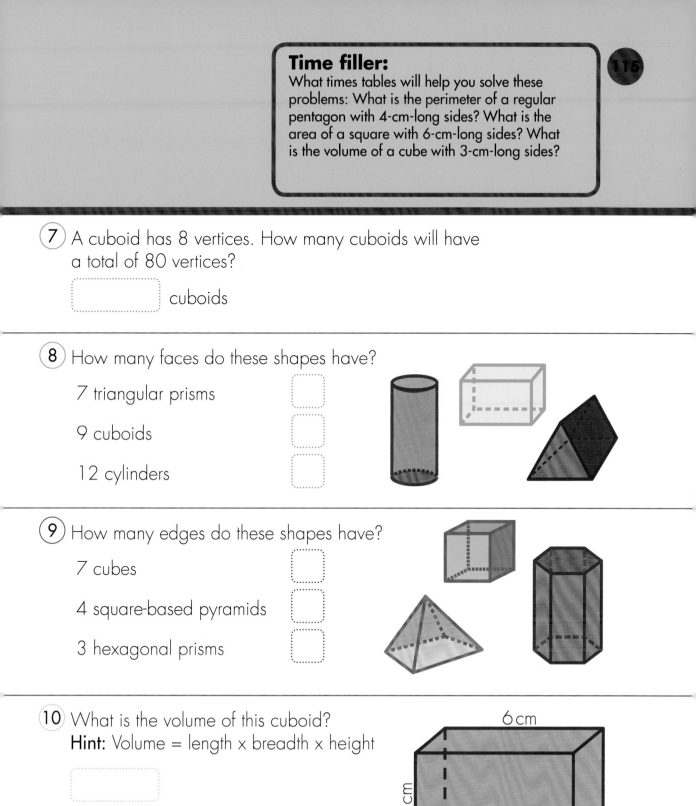

(9) How many edges do these shapes have?

7 cubes [   ]

4 square-based pyramids [   ]

3 hexagonal prisms [   ]

(10) What is the volume of this cuboid?
Hint: Volume = length x breadth x height

[   ]

6 cm

3 cm

4 cm

Are you ready to dive
Into counting in 5s?

(1) A pack of greeting cards contains 5 cards.
How many cards are there in 3 packs?

[   ] cards

(2) Complete each sequence:

0   5   10   [   ] [   ] [   ] [   ] [   ] [   ] [   ]

60   55   50   [   ] [   ] [   ] [   ] [   ] [   ] [   ]

75   80   85   [   ] [   ] [   ] [   ] [   ] [   ] [   ]

(3) Answer these questions:

Five groups of six are   [   ]

Seven multiplied by five is [   ]

Eleven times five is   [   ]

(4) David saved 24 5-pence
coins. How much money did
David save?

[   ]

(5) Work out these multiplication sums:

18	20	49	56	130	222
x 5	x 5	x 5	x 5	x 5	x 5

Time filler:
Calculate the total of the following amounts:

- 5 x 10p
- 5 x 5p
- 20 x 5p
- 25p ÷ 5

- 5 x 25p
- 15 x 5p
- 50 x 5p
- £1 ÷ 5

117

6 Divide each number by 5:

10 ☐　　　25 ☐　　　30 ☐　　　50 ☐　　　85 ☐

7 Five children are given £1.95 to share equally among them. How much money will each child receive?　☐

8 Work out these division sums:

☐　　　☐　　　☐　　　☐　　　☐

5$\overline{)65}$　　　5$\overline{)80}$　　　5$\overline{)125}$　　　5$\overline{)175}$　　　5$\overline{)250}$

9 There are 270 children in a school. There are 5 years, and each year has an equal number of children. How many children are there in Year 4?　☐ children

10 How many shapes are there in each group?

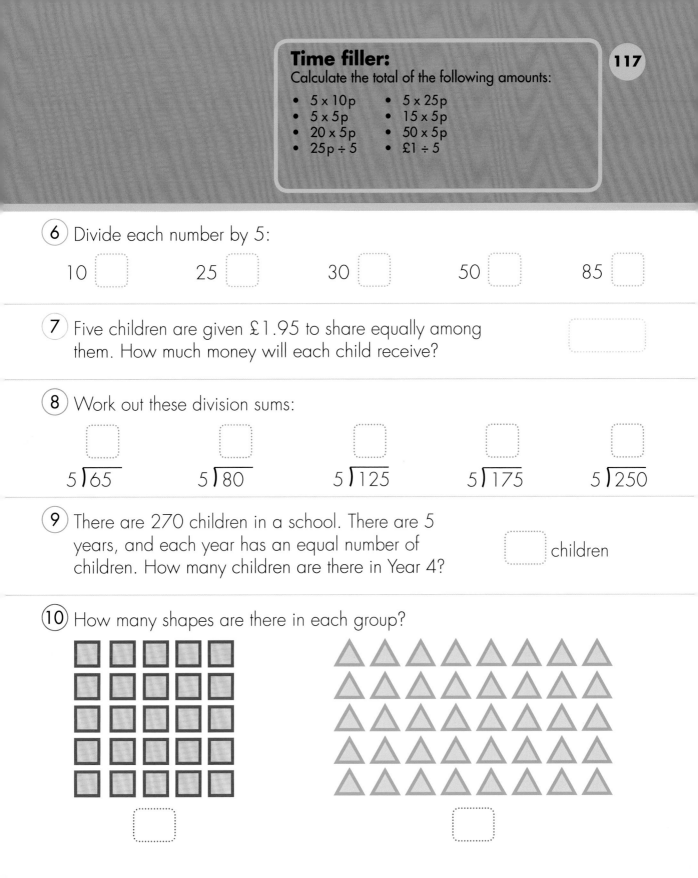

☐　　　　　　　　☐

Telling the time

Tick tock! Tick tock!
Ready to go? Then start the clock.

---

(1) How many minutes are there in 4 hours?

---

(2) How many minutes is it past 11 o'clock?

---

(3) How many minutes are there between 9.45 a.m. and 11.05 a.m.?

---

(4) How many minutes are there in 1 day?

---

(5) How many decades are there in half a century?
Note: A decade is 10 years; a century is 100 years.

decades

Time filler:
If a watch is 5 minutes fast, what is
the actual time if it reads:
9:55? 6:40? 2:20? 3:00?
If a watch is 5 minutes slow, what is
the actual time if it reads:
4:25? 7:00? 11:15? 12:00?

119

(6) How many hours are there in these months?

September (30 days) [   ]   February (28 days) [   ]

May (31 days) [   ]

(7) Write the number of minutes past the hour shown on each of these clocks.

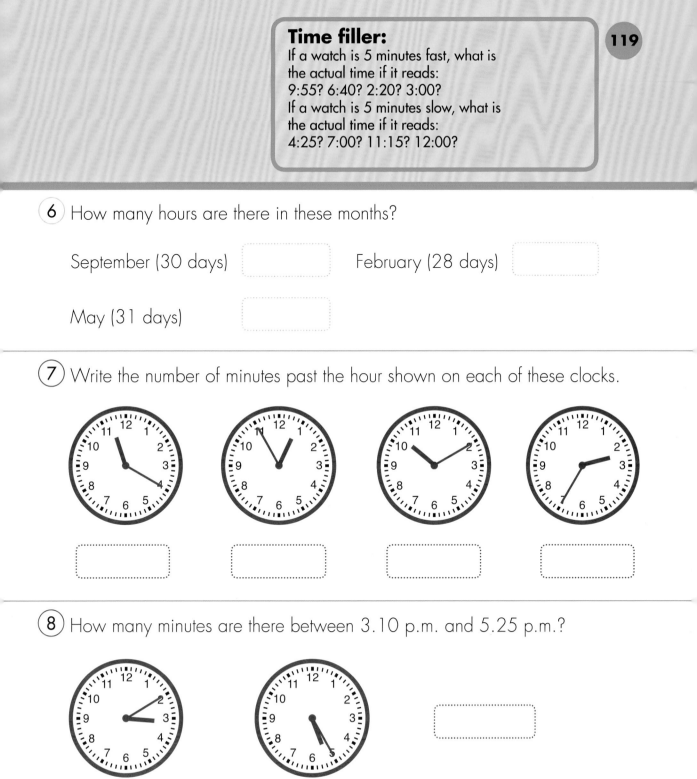

[   ]   [   ]   [   ]   [   ]

(8) How many minutes are there between 3.10 p.m. and 5.25 p.m.?

[   ]

120 Beat the clock 1

This is the place to gather pace.
How many answers do you know?
Get ready, get set, and go!

(1) 0 x 3 = ☐

(2) 12 x 2 = ☐

(3) 9 x 5 = ☐

(4) 1 x 4 = ☐

(5) 11 x 5 = ☐

(6) 3 x 4 = ☐

(7) 7 x 3 = ☐

(8) 10 x 2 = ☐

(9) 2 x 2 = ☐

(10) 6 x 5 = ☐

(11) 11 x 4 = ☐

(12) 8 x 2 = ☐

(13) 4 x 4 = ☐

(14) 10 x 5 = ☐

(15) 2 x 5 = ☐

(16) 6 x 3 = ☐

(17) 10 x 9 = ☐

(18) 9 x 4 = ☐

(19) 1 x 5 = ☐

(20) 12 x 4 = ☐

(21) 6 x 4 = ☐

(22) 7 x 4 = ☐

(23) 10 x 1 = ☐

(24) 12 x 10 = ☐

(25) 2 x 4 = ☐

(26) 10 x 7 = ☐

(27) 10 x 10 = ☐

(28) 8 x 4 = ☐

(29) 11 x 2 = ☐

(30) 0 x 2 = ☐

Time filler:
Check your answers on page 171. Return to this page again to improve your score.

121

(31) 24 ÷ 3 = ☐

(32) 6 ÷ 2 = ☐

(33) 15 ÷ 5 = ☐

(34) 30 ÷ 5 = ☐

(35) 9 ÷ 3 = ☐

(36) 30 ÷ 10 = ☐

(37) 14 ÷ 2 = ☐

(38) 3 ÷ 3 = ☐

(39) 50 ÷ 10 = ☐

(40) 40 ÷ 5 = ☐

(41) 2 ÷ 2 = ☐

(42) 90 ÷ 10 = ☐

(43) 21 ÷ 3 = ☐

(44) 6 ÷ 3 = ☐

(45) 10 ÷ 10 = ☐

(46) 18 ÷ 2 = ☐

(47) 27 ÷ 3 = ☐

(48) 80 ÷ 10 = ☐

(49) 15 ÷ 5 = ☐

(50) 15 ÷ 3 = ☐

(51) 60 ÷ 10 = ☐

(52) 40 ÷ 8 = ☐

(53) 36 ÷ 3 = ☐

(54) 70 ÷ 10 = ☐

(55) 18 ÷ 3 = ☐

(56) 60 ÷ 5 = ☐

(57) 110 ÷ 10 = ☐

(58) 30 ÷ 3 = ☐

(59) 25 ÷ 5 = ☐

(60) 100 ÷ 10 = ☐

122 Groups of 6

6, 12, 18, 24:
Count in 6s to make numbers more.

(1) A tube holds 6 tennis balls. How many balls will 8 tubes hold?

[ ] balls

(2) Complete each sequence:

0 6 12 [ ] [ ] [ ] [ ] [ ] [ ] [ ]

72 66 60 [ ] [ ] [ ] [ ] [ ] [ ] [ ]

54 60 66 [ ] [ ] [ ] [ ] [ ] [ ] [ ]

(3) Answer these questions:

Six threes are [ ]

Four multiplied by six is [ ]

Six groups of seven are [ ]

(4) Selma bought 9 bananas at 6p each. How much money did she spend?

[ ]

(5) Work out these multiplication sums:

14	25	38	54	116	200
x 6	x 6	x 6	x 6	x 6	x 6

Time filler:
Another way to work out the answer to 6x
a number is to multiply the number by 3
and then double the answer. Choose
some numbers between 1 and 20
and give this a go.

123

6 Divide each number by 6:

18 ☐   30 ☐   66 ☐   72 ☐   96 ☐

7 £2.50 is shared equally
among 6 children. How
much money is left over?

☐

8 174 cars are parked in 6 rows
of equal length. How many
cars are there in each row?

☐ cars

9 Work out these division sums:

$6\overline{)90}$ $6\overline{)102}$ $6\overline{)204}$ $6\overline{)276}$ $6\overline{)348}$

10 How many items are there in each group?

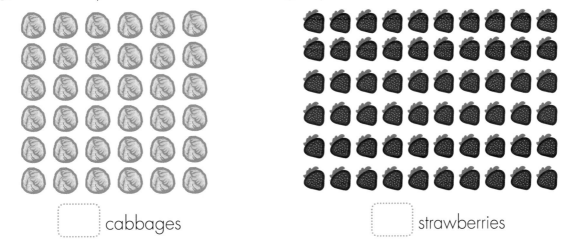

☐ cabbages   ☐ strawberries

124 Bugs

Multiply creepy-crawlies' legs
And then count up the butterfly eggs.

1 Will recorded the number of bugs he saw in his garden in a month.
How many legs did each type of insect have altogether?
Hint: An insect has 6 legs.

Name of bug	Number sighted	Total number of legs
Beetle	卌 卌 卌 II	
Wasp	卌 III	
Butterfly	卌 卌 卌	
Ladybird	卌 卌 卌 卌 I	

2 If a desert locust eats 2 g of food each day, how much will a swarm of 66 million desert locusts eat in a day?

3 A ladybird is about 6 mm long. Under the microscope, the ladybird is magnified 40 times. What size is the ladybird when seen through the microscope?

4 A leaf-cutter ant travels 360 m each day. Altogether, what is the total distance the ant will travel in 60 days?

Time filler:
A queen honey bee lays 1 500 eggs per day. How many eggs does she lay in 1 month (30 days)? How many eggs does she lay in 6 months (180 days)?

125

5 How many wings does a swarm of 6 000 bees have altogether?
Hint: Bees have 4 wings.

[_____] wings

6 A queen wasp can lay 2 000 eggs a day. How many eggs can she lay in 60 days?

[_____] eggs

7 A butterfly lays 600 eggs, and only a quarter of them hatch into caterpillars. How many eggs **do not** hatch?

[_____] eggs

8 Class 5F went to a pond. They made a picture chart of the number of bugs they saw in the pond. How many bugs of each type did they see?

● = 6 bugs   ◗ = 3 bugs

Name of bug	Number of bugs	Total
Pond skaters	○ ○ ○ ○	
Water bugs	○ ○ ○	
Whirligig beetles	○ ○ ◗	
Dragonfly nymphs	◗	
Water spiders	○	

Jump, throw, kick, and dash.
You will have this page done in a flash!

① Adam ran 400m in 59 seconds.
Jonas took twice as long.
How long did Jonas take?

② Three cyclists raced at 58mph, 63mph, and 56mph. What was their average speed?

③ These were the results of a season's football games:
A win = 5 points, a draw = 3 points, and a loss = 1 point.
How many points did each team get?

Football team	Win	Loss	Draw	Points
England	8	2	5	
United States	7	3	6	
Japan	7	1	7	
Brazil	9	2	4	

④ The winner of a tennis tournament won 4 times the prize money of a semi-finalist. If a semi-finalist received £475 000, how much money did the winner receive?

Time filler:
Seven competitors ran a 100 m race in a total time of 91 seconds. What was their average time?

127

(5) John threw a javelin a distance of 64 m. Amy threw the javelin an eighth ($\frac{1}{8}$) less than John's distance. How far did Amy throw?

> [ ]

(6) Seven race car drivers have a total of 1 645 points. What is the average number of points scored?

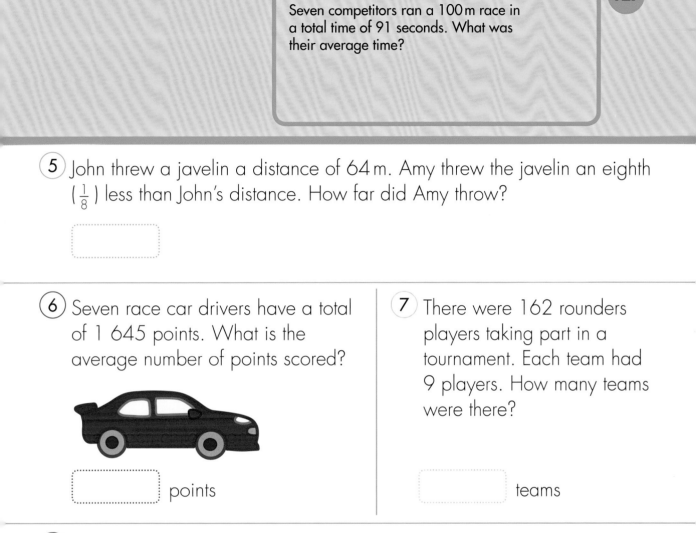

[ ] points

(7) There were 162 rounders players taking part in a tournament. Each team had 9 players. How many teams were there?

[ ] teams

(8) The length of a swimming pool is 25 m. This chart shows how many times each child swam that length. How far did each child swim?

Name of child	Number of lengths	Total distance
Harry	10	
Jasmine	8	
Jamie	6	
Heidi	5	

Groups of 7

7, 14, 21, 28:
Get started, do not wait!

① A dog eats 3 dog treats a day.
How many treats will it eat in 7 days?

:............: treats

② Complete each sequence:

0	7	14							
84	77	70							
35	42	49							

③ Answer these questions:

Seven sixes are :....:

Eight multiplied by seven is :....:

Five groups of seven are :....:

④ A train ticket costs £7.
How much will 6 tickets cost?

RAILWAYS

:............:

⑤ Work out these multiplication sums:

14	20	35	59	123	246
x 7	x 7	x 7	x 7	x 7	x 7
....	....	....	....	....	....

Time filler:
Another way to work out or check the answer to 7x a number is to multiply the number by 5, multiply the same number by 2 and then add the two answers. For example: $9 \times 7 = (9 \times 5) + (9 \times 2) = 45 + 18 = 63$. Choose a number between 1 and 20 and give this a go.

129

(6) Divide each number by 7:

0 ▢   21 ▢   49 ▢   77 ▢   98 ▢

(7) Seven books cost £35.84 altogether. If each book was the same price, what was the price of 1 book?   ▢

(8) Work out these division sums:

▢   ▢   ▢   ▢   ▢
7)84   7)140   7)105   7)133   7)224

(9) Share 42 chairs equally around 7 tables. How many chairs will you keep around each table?   ▢ chairs

(10) How many shapes are there in each group?

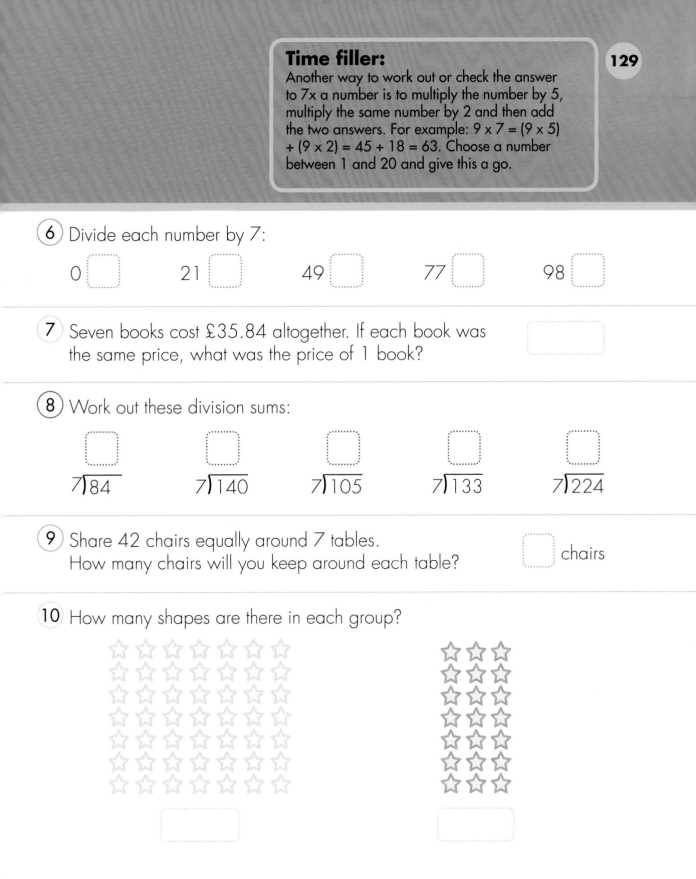

▢   ▢

Days of the week

Days 7; months 12; weeks 52;
Times tables are fun all year through.

1. How many days are there in 15 weeks?

   [_____] days

2. How many weeks are there in 7 years?

   [_____] weeks

3. How many hours are there in a week?

   [_____]

4. Dad works 35 hours a week. How many hours does he work over 4 weeks?

   [_____]

5. Fran cycles for 30 minutes every day. How many minutes does she cycle in one week?

   [_____]

6. A bookshop opens for 7 hours each day from Monday to Saturday. How many hours is it open in one week?

   BOOKSHOP

   [_____]

(7) Chris practises on the keyboard for 105 minutes every week.
He does an equal amount of time every day.
How long is each practice?

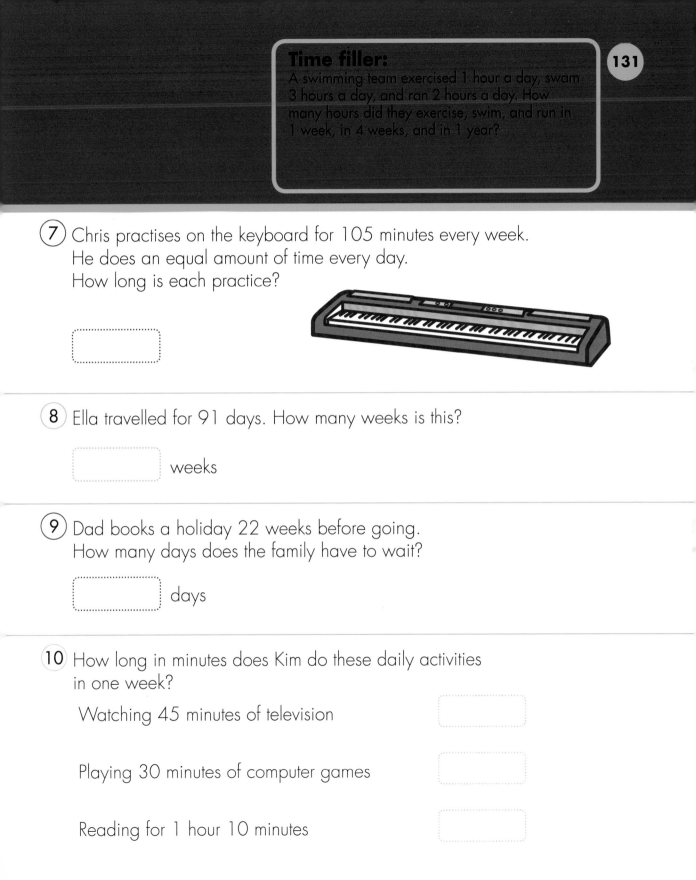

8 Ella travelled for 91 days. How many weeks is this?

_____ weeks

(9) Dad books a holiday 22 weeks before going.
How many days does the family have to wait?

_____ days

10 How long in minutes does Kim do these daily activities
in one week?

Watching 45 minutes of television

Playing 30 minutes of computer games

Reading for 1 hour 10 minutes

Dice and cards

Playing with dice and cards is fun
When your multiplication work is done.

1) Multiply the two numbers shown on the dice:

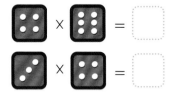 × = ⬚

× = ⬚

× = ⬚

× = ⬚

2) Add the numbers shown on the dice, and then multiply your answer by 6.

{ + } × 6 = ⬚

{ + } × 6 = ⬚

{ + } × 6 = ⬚

{ + } × 6 = ⬚

3) Jack threw a double six 5 times.
What was his total?

⬚

4) These are the scores of four players. They each need to throw a double to reach 100 points. What is the number that needs to appear on both dice for each player? Fill in the spaces in the table.

Player	Score	Number required on both dice
1	90	
2	98	
3	94	
4	88	

Time filler:
Calculate these answers:
- the product of 7 diamonds, 5 diamonds, and 4 diamonds;
- the product of 8 spades, 6 spades, and 2 spades;
- the product of 5 hearts, 9 hearts, and 3 hearts.

133

⑤ Jess throws a die 100 times and records her scores. What is the total amount scored altogether by Jess? Fill in the spaces in the table.

Number on die	Number of times thrown	Total
1	卌 卌 卌 |	
2	卌 卌 卌 卌	
3	卌 卌 卌 |||	
4	卌 卌 卌	
5	卌 卌 卌 ||	
6	卌 卌 ||||	
	Total	

⑥ A full deck of cards has 13 cards of each suit.

Note: There are 4 suits in a deck.

How many cards are there in a deck? ⬜ cards

How many cards are there in

4 decks? ⬜ cards

9 decks? ⬜ cards

6 decks? ⬜ cards

Multiply each of these cards by 8:

9 hearts ⬜

8 diamonds ⬜

Queen (12) clubs ⬜

Groups of 8

Get started now and do not be late!
Make steady progress as you count in 8s.

---

(1) Six trains run every hour. Each train pulls 8 carriages.
How many carriages are pulled every hour?

[　　　　] carriages

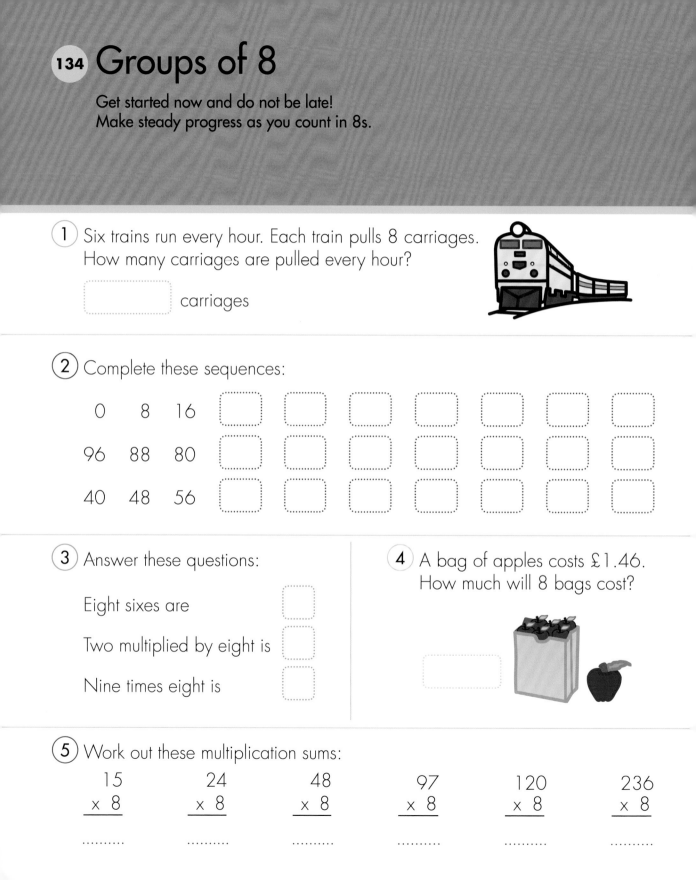

---

(2) Complete these sequences:

0	8	16	[　]	[　]	[　]	[　]	[　]	[　]	[　]
96	88	80	[　]	[　]	[　]	[　]	[　]	[　]	[　]
40	48	56	[　]	[　]	[　]	[　]	[　]	[　]	[　]

---

(3) Answer these questions:

Eight sixes are [　]

Two multiplied by eight is [　]

Nine times eight is [　]

(4) A bag of apples costs £1.46.
How much will 8 bags cost?

[　　　　]

---

(5) Work out these multiplication sums:

15	24	48	97	120	236
× 8	× 8	× 8	× 8	× 8	× 8
....	....	....	....	....	....

Time filler:
Another way to work out the answer to 8x a number is to multiply the number by 4 and then double the answer; or double the number, double the answer, and double again. Choose some numbers between 1 and 20 and give both these ways a go.

6 Work out these division sums:

$8\overline{)96}$ $8\overline{)144}$ $8\overline{)168}$ $8\overline{)256}$ $8\overline{)312}$

7 Tammy needs 192 m of fencing to go around her garden. Each fencing panel is 8 m long. How many panels will she need?

_____ panels

8 Perry pays £2.80 for 8 pencils. How much did 1 pencil cost?

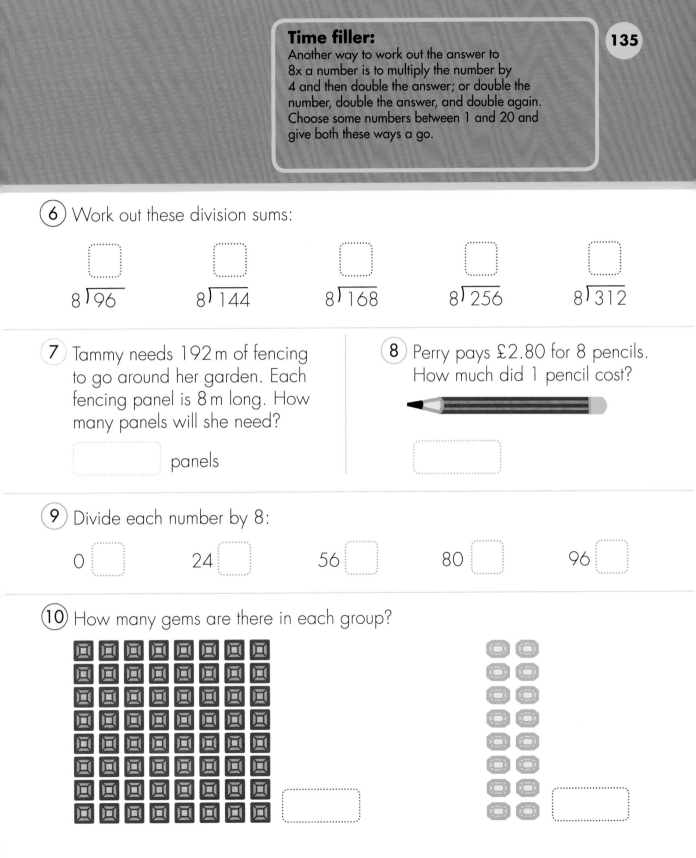

9 Divide each number by 8:

0 [ ]   24 [ ]   56 [ ]   80 [ ]   96 [ ]

10 How many gems are there in each group?

Solar system

5, 4, 3, 2, 1;
Blast off into space to get these done.

1. Thirty-five satellites are launched into orbit each year.
How many satellites have been launched in 8 years?

   [ ] satellites

2. A probe travelling at 500 mph will take 8 years to get
to Mars. How fast does the probe need to travel to get
there in 1 year?

   [ ]

3. Mars takes approximately 687 days to orbit the Sun.
How many days will it take for Mars to orbit the Sun 8 times?

   [ ] days

4. Neptune takes 165 years to orbit the Sun. How many
years will it take to go around the Sun 8 times?

   [ ] years

5. Mercury takes 88 Earth days to orbit the Sun. How many times
will Mercury go around the Sun in 880 Earth days?

   [ ] times

Time filler:
A team of 7 astronauts prepare for an 8-day space mission. If they each need 1.7 kg of food per day and 2 litres of water, what is the minimum amount they need to take with them?

(6) Multiply the 8 moons of Neptune with the 16 moons of Jupiter.

.... moons

(7) The midday surface temperature of Mercury is 420°C.
What is 8 times hotter than the surface of Mercury?

(8) Mercury is 4 878 km in diameter. What is an eighth ($\frac{1}{8}$) of its diameter?

(9) One day on Saturn lasts 10 hours and 14 minutes on Earth.
How long in Earth time will 8 days on Saturn last?

One day on Uranus lasts 17 hours and 8 minutes on Earth.
How long in Earth time will 8 days on Saturn last?

(10) Saturn is 1.4 billion km away from the Sun.
What is $\frac{5}{8}$ of this distance?

Jupiter is 780 million km away from the Sun.
What is $\frac{3}{8}$ of this distance?

138 Fractions

Split the whole by the denominator,
Then multiply by the numerator.

1. What is half ($\frac{1}{2}$) of each number?

   18 [ ]   10 [ ]   6 [ ]   24 [ ]

2. What is a third ($\frac{1}{3}$) of each amount?

   12g [ ]   27g [ ]   33g [ ]   42g [ ]

3. What is a quarter ($\frac{1}{4}$) of each number?

   4 [ ]   20 [ ]   36 [ ]   52 [ ]

4. There are 60 carrots in a box.
   How many carrots make up...

   $\frac{7}{10}$ of the box? [ ] carrots

   $\frac{1}{10}$ of the box? [ ] carrots $\frac{2}{10}$ of the box? [ ] carrots

5. There were 25 bananas, and $\frac{1}{5}$ were eaten.
   How many bananas are left?

   [ ] bananas

Time filler:
Calculate the following amounts:
$\frac{5}{9}$ of 450 g; $\frac{7}{10}$ of £2.50; $\frac{3}{8}$ of 640 cm.
Write some fraction challenges for your
friends. Did they get them right?

6 What is three quarters ($\frac{3}{4}$) of each number?

12 ☐   24 ☐   32 ☐   44 ☐

7 What is $\frac{1}{8}$ of 48 slices of pizza?

☐ slices

8 What is $\frac{7}{10}$ of 40?

☐

9 There are 30 children in a class. $\frac{3}{5}$ of the class have school dinners.
How many children **do not** have school dinners?

☐ children

10 Oliver picked 54 apples. $\frac{1}{6}$ were rotten.
How many apples were rotten?

☐ apples

Groups of 9

Wide awake and ready to shine?
Here is counting in sets of nine.

---

(1) There are 8 horse races in a day. If 9 different horses took part in each race, how many horses ran that day?

[ ] horses

---

(2) Complete each sequence:

0   9   18   [ ] [ ] [ ] [ ] [ ] [ ] [ ]

108   99   90   [ ] [ ] [ ] [ ] [ ] [ ] [ ]

45   54   63   [ ] [ ] [ ] [ ] [ ] [ ] [ ]

---

(3) Answer these questions:

Three multiplied by nine is   [ ]

Nine eights are   [ ]

Six groups of nine are   [ ]

(4) A bunch of flowers costs £4.99. How much will 9 bunches cost?

[ ]

---

(5) Work out these multiplication sums:

16	23	92	47	150	218
x 9	x 9	x 9	x 9	x 9	x 9
....	....	....	....	....	....

(6) Divide each number by 9:

9 [ ]   36 [ ]   45 [ ]   90 [ ]   108 [ ]

(7) Jake needed £468 to buy a new television.
He decided to save an equal amount over
9 weeks to reach the total. What is the
amount he needed to save each week?

[ ]

(8) How many shapes are there in each group?

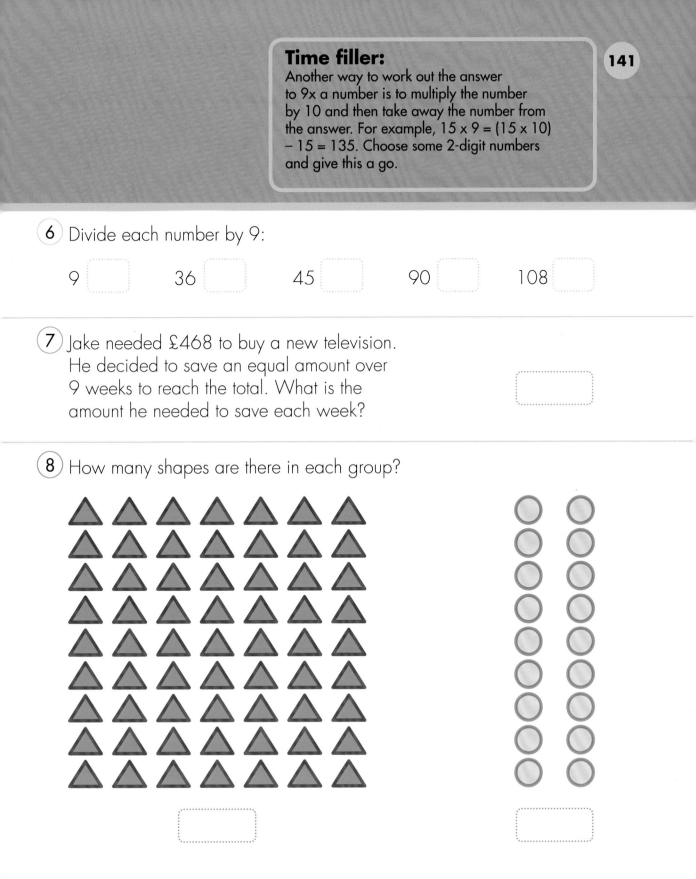

[ ]   [ ]

Shopping

Pick, choose, and weigh;
Then go to the checkout to pay!

① Calculate the total cost that Karl spent shopping.

Item	Cost per item	Amount	Total
Tomatoes	20p	6	
Carrots	10p	8	
Cabbage	89p	2	
Peppers	43p	5	
Cheese	£1.26	1	
Bread	76p	3	
Juice	£1.49	4	
Milk	72p	6	
Biscuits	89p	7	
Pasta	£2.56	2	

Time filler:
Calculate the total of these amounts:
- 3 kg of apples at 45p a kilogram;
- 7 oranges at 23p each;
- 6 cans of soup at £1.20 a can.
  Which is cheaper?
  12 packets of biscuits at 54p each
  or 9 jars of jam at 63p each?

143

(2) A shopkeeper sold 6 red coats at £89 each.
How much did the red coats cost altogether?

(3) Mum bought 3 bracelets at £7.84 each.
How much money did Mum spend?

(4) Tami spent £77.97 on 3 pairs of shoes. Each pair cost the same
amount. How much did each pair cost?

(5) In a sale, the cost of a hat was reduced by 20%. The original price of
the hat was £14.50. How much was it reduced by?

(144) # Beat the clock 2

This is the place to gather pace.
How many answers do you know?
Get ready, get set, and go!

(1) $0 \times 9 =$

(2) $0 \times 6 =$

(3) $8 \times 9 =$

(4) $5 \times 8 =$

(5) $7 \times 8 =$

(6) $3 \times 8 =$

(7) $2 \times 9 =$

(8) $7 \times 6 =$

(9) $2 \times 7 =$

(10) $1 \times 7 =$

(11) $9 \times 9 =$

(12) $10 \times 9 =$

(13) $2 \times 8 =$

(14) $1 \times 8 =$

(15) $12 \times 6 =$

(16) $3 \times 6 =$

(17) $9 \times 7 =$

(18) $10 \times 7 =$

(19) $4 \times 7 =$

(20) $4 \times 9 =$

(21) $11 \times 6 =$

(22) $8 \times 8 =$

(23) $6 \times 8 =$

(24) $11 \times 8 =$

(25) $9 \times 6 =$

(26) $6 \times 7 =$

(27) $12 \times 7 =$

(28) $5 \times 6 =$

(29) $7 \times 7 =$

(30) $12 \times 9 =$

Time filler:
Check your answers on page 171.
Return to this page again to improve your score.

145

(31) $0 \div 7 =$ ☐

(32) $60 \div 6 =$ ☐

(33) $63 \div 9 =$ ☐

(34) $0 \div 8 =$ ☐

(35) $99 \div 9 =$ ☐

(36) $84 \div 7 =$ ☐

(37) $9 \div 9 =$ ☐

(38) $27 \div 9 =$ ☐

(39) $48 \div 8 =$ ☐

(40) $6 \div 6 =$ ☐

(41) $96 \div 8 =$ ☐

(42) $72 \div 8 =$ ☐

(43) $21 \div 7 =$ ☐

(44) $32 \div 8 =$ ☐

(45) $63 \div 7 =$ ☐

(46) $24 \div 6 =$ ☐

(47) $45 \div 9 =$ ☐

(48) $35 \div 7 =$ ☐

(49) $77 \div 7 =$ ☐

(50) $48 \div 6 =$ ☐

(51) $54 \div 9 =$ ☐

(52) $80 \div 8 =$ ☐

(53) $49 \div 7 =$ ☐

(54) $12 \div 6 =$ ☐

(55) $36 \div 6 =$ ☐

(56) $64 \div 8 =$ ☐

(57) $56 \div 7 =$ ☐

(58) $40 \div 8 =$ ☐

(59) $81 \div 9 =$ ☐

(60) $108 \div 9 =$ ☐

Division

Use your times tables know-how
To work out division questions now.

---

(1) Match each question to its answer:

168 ÷ 6   524 ÷ 4   595 ÷ 7   729 ÷ 9

85   81   28   131

---

(2) Use the long division method to work out each sum:

☐   ☐   ☐   ☐

4⟌648   2⟌496   5⟌760   6⟌822

---

(3) What is the remainder each time?

592 ÷ 3 ☐   264 ÷ 7 ☐

786 ÷ 4 ☐   543 ÷ 9 ☐

---

(4) Circle all the multiples of 7:

14   23   35   43   76   84

Time filler:
How many multiples of 9 are also the answer to other times tables? For example, 18 is a multiple of 9 and is also the answer to 1 x 18, 2 x 9 and 3 x 6. Make a list of all the overlaps.

---

(5) Circle all the multiples of 9:

28   54   61   83   99   108

---

(6) Circle all the multiples of 12:

24   45   60   56   72   98   132

---

(7) Work out these money sums:

£14.58 ÷ 3 = ⬚   £35.60 ÷ 8 = ⬚

£26.96 ÷ 4 = ⬚   £66.69 ÷ 9 = ⬚

---

(8) List all the factors for each number:

24 ⬚ ⬚ ⬚ ⬚ ⬚ ⬚ ⬚ ⬚

36 ⬚ ⬚ ⬚ ⬚ ⬚ ⬚ ⬚ ⬚ ⬚

72 ⬚ ⬚ ⬚ ⬚ ⬚ ⬚ ⬚ ⬚ ⬚ ⬚ ⬚ ⬚

100 ⬚ ⬚ ⬚ ⬚ ⬚ ⬚ ⬚ ⬚ ⬚

Groups of 11

11, 22, 33, 44;
Follow the pattern to find more.

1. A farmer plants 6 rows of tulips, with 11 bulbs in each row. How many tulip bulbs are planted?

   [   ] bulbs

2. Complete each sequence:

   0   11   22   [ ] [ ] [ ] [ ] [ ] [ ] [ ]

   143 132 121 [ ] [ ] [ ] [ ] [ ] [ ] [ ]

   66   77   88   [ ] [ ] [ ] [ ] [ ] [ ]

3. Answer these questions:

   Eleven fours are   [ ]

   Eleven groups of seven are   [ ]

   Twelve times eleven is   [ ]

4. Ellie buys 11 T-shirts at £1.10 each. How much does Ellie pay for 11 T-shirts?

   [   ]

5. Work out these multiplication sums:

14	25	69	33	81	100
x 11	x 11	x 11	x 11	x 11	x 11

Time filler:
Another way to work out or check the answer
to 11x a number is to multiply the number by
10 and then add on the number to the answer.
For example, 23 x 11 = (23 x 10) + 23 = 253.
Choose some 2-digit numbers and give this a go.

6 Divide each number by 11:

22 ☐   88 ☐   121 ☐   143 ☐   176 ☐

7 Work out these division sums:

$11\overline{)187}$ $11\overline{)297}$ $11\overline{)363}$ $11\overline{)572}$ $11\overline{)781}$

8 How many spots are there in each group?

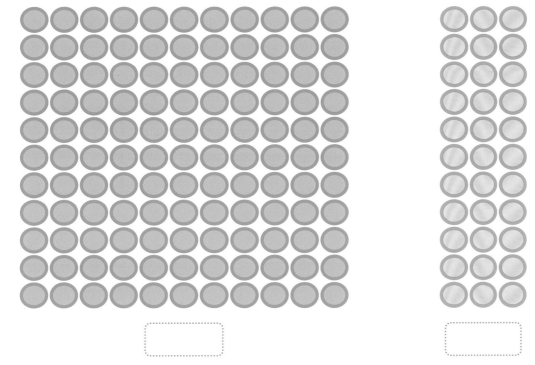

☐   ☐

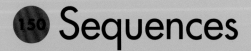

Sequences

Find the number pattern for each row;
The times tables facts are fun to know.

(1) Fill in the missing numbers in each sequence:

5   10   ☐   ☐   ☐   30   ☐   ☐   45   ☐

60   ☐   ☐   75   ☐   ☐   ☐   95   ☐   105

(2) Complete each sequence:

4   8   12   ☐   ☐   ☐   ☐   ☐   ☐   ☐

7   14   21   ☐   ☐   ☐   ☐   ☐   ☐   ☐

25   50   75   ☐   ☐   ☐   ☐   ☐   ☐   ☐

(3) Continue this pattern:

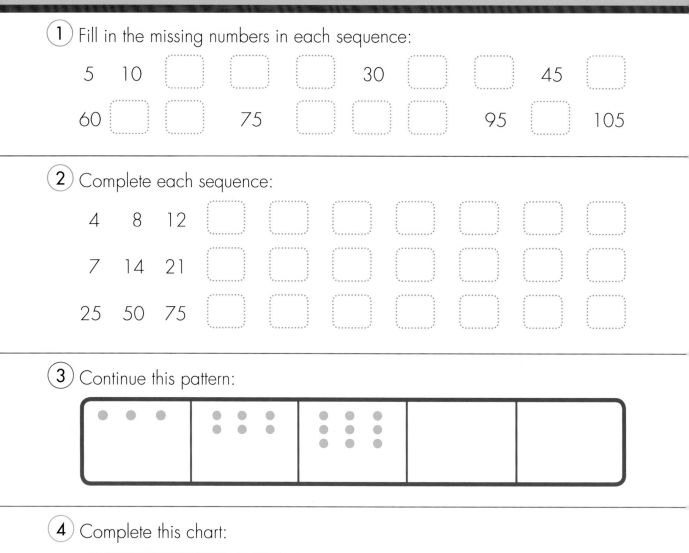

(4) Complete this chart:

×	0	1	2	3	4	5	6	7	8	9	10
6				18							
9						45					

Time filler:
- Can you work out the numbers in these sequences:
- start on 3 and multiply by 3 for 6 steps?
- start on 8 and double the number for 5 steps?

(5) Fill in the missing numbers in each sequence:

80 72 ☐ ☐ ☐ 40 ☐ ☐ ☐ 8

60 56 52 ☐ ☐ ☐ 36 ☐ ☐ ☐

(6) Continue this pattern:

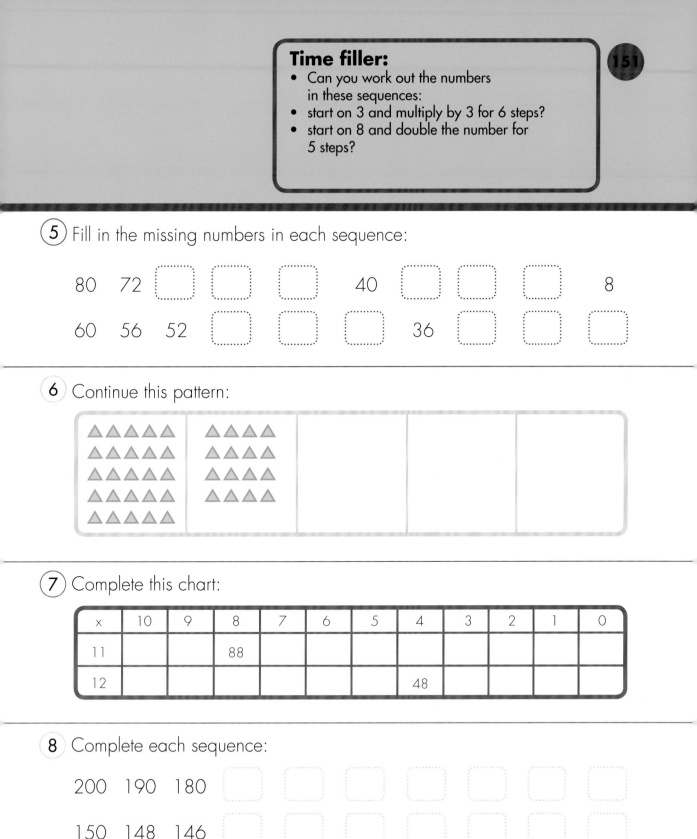

(7) Complete this chart:

x	10	9	8	7	6	5	4	3	2	1	0
11			88								
12							48				

(8) Complete each sequence:

200 190 180 ☐ ☐ ☐ ☐ ☐ ☐

150 148 146 ☐ ☐ ☐ ☐ ☐ ☐

Groups of 12

Here is the last of the times tables;
You know most of these.
They have been in the others,
So you can answer with ease.

(1) A baker takes an hour to cook 12 loaves of bread.
How many loaves can he make in 3 hours?

☐ loaves

(2) Complete each sequence:

0   12   24   ☐ ☐ ☐ ☐ ☐ ☐ ☐

144 132 120   ☐ ☐ ☐ ☐ ☐ ☐ ☐

60   72   84   ☐ ☐ ☐ ☐ ☐ ☐ ☐

(3) Answer these questions:

Five groups of twelve are ☐

Eight times twelve is ☐

Ten multiplied by twelve is ☐

(4) Cara collects trading cards. She buys 12 packs of 4 cards at £2 per pack. How much money does Cara spend and how many trading cards will she have?

☐   ☐ trading cards

(5) Work out these multiplication sums.

13	17	24	35	42	100
×12	×12	×12	×12	×12	×12
....	....	....	....	....	....

Time filler:
Another way to work out or check the answer to 12x a number is to multiply the number by 10, multiply the number by 2, and then add the two answers. For example: 14 x 12 = (14 x 10) + (14 x 2) = 140 + 28 = 168. Choose some 2-digit numbers and give this a go.

153

6 Divide each number by 12:

0 [ ]   48 [ ]   84 [ ]   132 [ ]   192 [ ]

7 Mum has a loan of £864, which she pays back in equal amounts over 12 months. How much does Mum pay each month?

[ ]

8 How many shapes are there in each group?

[ ]   [ ]

Dozen a day

Keep multiplying by 12 to find the way
To get the answers for a Dozen a day.

① A dozen children split themselves equally into 3 teams to play a game.
How many children are there in each team?

.... children

② Cupcakes were sold in boxes of 12.
How many cupcakes were there in 15 boxes?

.... cupcakes

③ How many dozen eggs are there in
a gross of eggs? **Hint:** A gross is 144.

....

④ What are the factors of 12?

....

⑤ A score of children had 12 sweets each.
How many sweets did they have altogether?
Note: A score is 20.

.... sweets

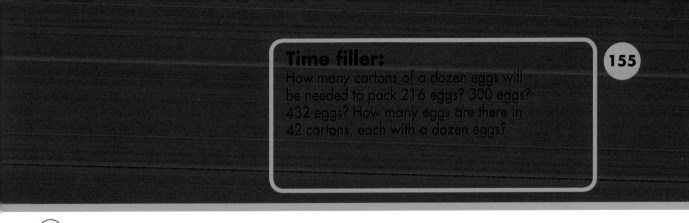

Time filler:
How many cartons of a dozen eggs will
be needed to pack 216 eggs? 300 eggs?
432 eggs? How many eggs are there in
42 cartons, each with a dozen eggs?

155

6) A chef cooks a batch of 12 pancakes in 12 minutes.
How many batches of 12 pancakes can he make in an hour?

[_____] batches

7) Trains arrived at Whistlestop
Station 3 times an hour.
How many trains arrived
in 12 hours?

[_____] trains

8) A group of musicians performed
a dozen pieces in a concert.
Each piece lasted 4 minutes.
How many minutes did the
musicians perform altogether?

[_____]

9) 900 raffle tickets were sold at a
fund-raising event. There were a dozen
prizes. What was the chance of winning
a prize? Circle the correct answer.

1 in 50   1 in 75   1 in 100

10) Once a month, Jill ran a distance of 10 km in a cross-country
event. How many kilometres did Jill run in a year?

[_____]

Beat the clock 3

This is the place to gather pace.
How many answers do you know?
Get ready, get set, and go!

(1) 3 x 9 = ☐

(2) 19 x 4= ☐

(3) 20 x 6 = ☐

(4) 1 x 7 = ☐

(5) 12 x 9= ☐

(6) 12 x 8 = ☐

(7) 4 x 6 = ☐

(8) 10 x 8= ☐

(9) 12 x 5 = ☐

(10) 5 x 4 = ☐

(11) 11 x 0= ☐

(12) 11 x 2 = ☐

(13) 5 x 8 = ☐

(14) 11 x 6= ☐

(15) 18 x 3 = ☐

(16) 8 x 2 = ☐

(17) 12 x 4= ☐

(18) 16 x 2 = ☐

(19) 7 x 7 = ☐

(20) 10 x 5= ☐

(21) 14 x 0 = ☐

(22) 7 x 3 = ☐

(23) 15 x 9= ☐

(24) 32 x 1= ☐

(25) 1 x 3 = ☐

(26) 17 x 8= ☐

(27) 19 x 10 = ☐

(28) 2 x 10 = ☐

(29) 15 x 7= ☐

(30) 16 x 12 = ☐

Time filler:
Check your answers on page 172. Return to this page again to improve your score.

157

(31) 54 ÷ 9 = [ ]

(32) 245 ÷ 5 = [ ]

(33) 85 ÷ 5 = [ ]

(34) 81 ÷ 9 = [ ]

(35) 112 ÷ 8 = [ ]

(36) 24 ÷ 3 = [ ]

(37) 56 ÷ 8 = [ ]

(38) 100 ÷ 5 = [ ]

(39) 92 ÷ 2 = [ ]

(40) 56 ÷ 7 = [ ]

(41) 108 ÷ 6 = [ ]

(42) 63 ÷ 3 = [ ]

(43) 42 ÷ 6 = [ ]

(44) 456 ÷ 1 = [ ]

(45) 28 ÷ 2 = [ ]

(46) 91 ÷ 7 = [ ]

(47) 537 ÷ 3 = [ ]

(48) 76 ÷ 2 = [ ]

(49) 40 ÷ 4 = [ ]

(50) 860 ÷ 10 = [ ]

(51) 99 ÷ 11 = [ ]

(52) 25 ÷ 5 = [ ]

(53) 240 ÷ 12 = [ ]

(54) 320 ÷ 10 = [ ]

(55) 51 ÷ 3 = [ ]

(56) 143 ÷ 11 = [ ]

(57) 144 ÷ 12 = [ ]

(58) 64 ÷ 4 = [ ]

(59) 121 ÷ 11 = [ ]

(60) 1000 ÷ 10 = [ ]

100–101 Groups of 2
102–103 Pairs and doubles

100

1. Anne has 2 baskets with 5 flowers in each basket. How many flowers are there in total? Fill in the numbers.

   [2] baskets x [5] flowers = [10] flowers

2. Complete each sequence:

   2 4 6 [8] [10] [12] 14 [16] [18] [20] 22 [24]

   48 46 44 [42] [40] [38] [36] [34] 32 [30] [28] 26

   54 56 58 [60] [62] 64 [66] [68] 70 [72] 74 [76]

3. Answer these questions:
   Six multiplied by two is [12]
   Seven times two is [14]
   Nine groups of two are [18]

4. A theatre ticket costs £24.50. How much will 2 tickets cost?

   [£49.00]

5. Work out these multiplication sums:

150	175	236	348	427	519
x 2	x 2	x 2	x 2	x 2	x 2
300	350	472	696	854	1 038

101

6. Divide each number by 2:

   76 [38]   142 [71]   178 [89]

7. Work out these division sums:

   [63]   [120]   [176]   [342]   [396]
   2)126   2)240   2)352   2)684   2)792

8. Fazir and Tira shared £7.80 equally between them. How much money did each child receive?

   [£3.90]

9. There were 284 bees in 2 hives. If there was an equal number in each hive, how many bees were there in 1 hive?

   [142] bees

10. How many shapes are there in each group?

    [8]   [16]

All the pages in this chapter are intended for children who are familiar with the 0 to 12 times tables and are able to work out long multiplication and division sums. At this age, your child will know that multiplication is a fast way of adding equal groups of numbers.

102

1. Double each number:

   25 [50]   42 [84]   70 [140]   127 [254]

2. How many socks are there in 36 pairs?

   [72] socks

3. In a day, a factory makes wheels for 350 bicycles. How many wheels are made in 1 day?

   [700] wheels

4. The table below shows some ingredients needed to make 12 biscuits. Calculate how much of each you will need to make 24 biscuits.
   Hint: Double each amount.

Ingredients	Quantity for 12 biscuits	Quantity for 24 biscuits
Flour	350g (12oz)	700g (24oz)
Eggs	2	4
Butter	225g (8oz)	450g (16oz)
Caster sugar	175g (6oz)	350g (12oz)
Dark chocolate	350g (12oz)	700g (24oz)
Light brown sugar	175g (6oz)	350g (12oz)

103

5. How many wings do 275 crane flies have altogether?
   Note: A crane fly has two wings.

   [550] wings

6. The chart shows the number of bunches of flowers sold in a store in one week. Write the total for each day.

   = 2 bunches

Day	Number of bunches sold	Total
Monday	10 x	20
Tuesday	8 x	16
Wednesday	12 x	24
Thursday	9 x	18
Friday	20 x	40
Saturday	14 x	28
Sunday	5 x	10

7. Ryan cycles 56 km, but Jake cycles twice as far. How far does Jake cycle?

   112km

8. Mum has spent £84.00 on presents for Jayden, but Dad has spent double that amount. How much money has Dad spent?

   [£168.00]

Children will be ready for challenging questions to apply their times tables knowledge. They will learn to read the question carefully, identify the data, and decide which of the four operations to use, which will be either to multiply or to divide in this chapter.

Answers:

104–105 Groups of 10
106–107 Multiplying by 100 and 1 000

104

① Iiya had 7 parcels. Each parcel weighed 10 kg. How much did the parcels weigh altogether?

70 kg

② Complete these sequences:

10 20 30 | 40 | 50 | 60 | 70 | 80 | 90 | 100 |

150 140 130 | 120 | 110 | 100 | 90 | 80 | 70 | 60 |

270 280 290 | 300 | 310 | 320 | 330 | 340 | 350 | 360 |

③ Answer these questions:

Ten eights are | 80 |

Ten times ten is | 100 |

Nine multiplied by ten is | 90 |

④ Zina saved 35 10-pence coins. How much money did Zina have altogether?

£3.50

⑤ Work out these multiplication sums:

436	845	152	1 689	791	287
×10	×10	×10	×10	×10	×10
4 360	8 450	1 520	16 890	7 910	2 870

105

⑥ Divide each number by 10:

10 | 1 | 40 | 4 | 80 | 8 | 120 | 12 | 150 | 15 |

⑦ Work out these division sums:

| 42 | 36.7 | 78 | 84.2 | 99 |
| 10)420 | 10)367 | 10)780 | 10)842 | 10)990 |

⑧ How many leaves are there in each group? **Hint:** Multiply the number of rows by the number of columns.

80   70

Children will know that multiplying by 10 means to make amounts ten times bigger. In the case of decimals, this means moving the decimal point one place to the right to multiply and moving the decimal point one place to the left to divide.

106

① Multiply each number by 100:

4 | 400 | 47 | 4 700 | 470 | 47 000 | 4 070 | 407 000 |

② A box contains 100 T-shirts. How many T-shirts are there in 64 boxes?

6 400 T-shirts

③ Multiply the number by 100 each time:

| 3 | 300 | 30 000 | 3 000 000 |
| 82 | 8 200 | 820 000 | 82 000 000 |

④ How many centimetres are there in 84 m?   8 400 cm

⑤ Divide each number by 100:

42 000 | 420 | 702 000 | 7 020 |

804 200 | 8 042 | 6 000 000 | 60 000 |

107

⑥ Multiply each number by 1 000:

7 | 7 000 | 82 | 82 000 | 146 | 146 000 | 150 | 150 000 |

⑦ How many grams are there in 7.2 kg?

7 200 g

⑧ How many pence are there in £35?

3 500 p

⑨ A plane flies at a height of 10 668 m. What is this height in kilometres?

10.668 km

⑩ A colony of army ants has 700 000 ants. As the ants cross a river, 20% of the colony dies. How many ants make it across?

560 000 ants

Children will be familiar with the equivalents between metric units of measurement and that metric is based on multiples of 10. Some of these questions require the knowledge that 100 cm = 1 m, 1 000 m = 1 km, and 1 000 g = 1 kg. Also, dividing by 100 helps solve percentage problems, as percent means part per 100.

Answers:

108–109 Groups of 3
110–111 Triple fun

① A jar holds 8 biscuits. How many biscuits are there in 3 jars?

| 24 | biscuits |

② Complete each sequence:

0 3 6 [9] [12] 15 [18] 21 [24] 27

36 33 30 [27] [24] 21 [18] 15 [12] 9

36 39 42 [45] [48] 51 [54] 57 60 63

③ Answer these questions:

Three fives are — 15

Three multiplied by seven is — 21

Three times nine is — 27

④ Neo bought 6 oranges at 30p each. What was the total cost of the 6 oranges?

| £1.80 |

⑤ Work out these multiplication sums:

16	33	55	79	145	229
× 3	× 3	× 3	× 3	× 3	× 3
48	99	165	237	435	687

⑥ Divide each number by 3:

6 [2]   15 [5]   24 [8]   36 [12]   45 [15]

⑦ How long will it take Anita to save 42p if she saves 3p every week? | 14 | weeks

⑧ Work out these division sums:

| [20] | [30] | [24] | [33] | [61] |
| 3)60 | 3)90 | 3)72 | 3)99 | 3)183 |

⑨ Pablo was paid £3 for each car that he washed. He earned £39 in one week. How many cars did Pablo wash that week?

| 13 | cars

⑩ How many shapes are there in each group?

| 21 | | 27 |

These orange pages focus on a particular times table and provide a wide range of questions to reinforce your child's familiarity with the specific times table. The sequences support their knowledge of the multiples, which is very useful to know when solving division sums.

① How many wheels are there on 15 tricycles?

| 45 | wheels

② How many sides do 55 triangles have?

| 165 | sides

③ About 170 triplets are born in the United Kingdom each year. How many babies is this?

| 510 | babies

④ Thirty-nine trimarans race in a competition. How many hulls are there altogether? **Note:** A trimaran is a boat with 3 hulls.

| 117 | hulls

⑤ Fifty-four children are split into groups of 3. How many groups of children are there?

| 18 | groups

⑥ A magnifying glass makes bugs look triple their size. Below are the original sizes of the bugs. What size is each of the bugs when it is magnified?

Worm: 6.5cm | 19.5cm |

Centipede: 5.25cm | 15.75cm |

Ladybird: 1.75cm | 5.25cm |

⑦ Packets of biscuits are sold in boxes of 3 packets. This chart shows how many boxes are sold from a store in a week. Calculate the number of packets sold that week.   = 3 packets

Day	Number of boxes	Total
Monday		6
Tuesday		15
Wednesday		12
Thursday		24
Friday		3
Saturday		27
Sunday		21

⑧ Leaving no spaces, fit 9 small triangles (of equal size) inside the large equilateral triangle.

These red pages are themed and based on a particular times table. Times table knowledge is relevant and can be useful in many everyday activities and situations. Hopefully, children can be motivated and will understand that multiplication and division facts are applicable to life.

Answers:

112–113 Groups of 4
114–115 Shapes

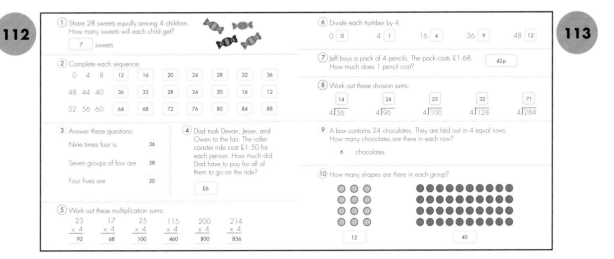

112

① Share 28 sweets equally among 4 children. How many sweets will each child get?

7 sweets

② Complete each sequence:

0 4 8 12 16 20 24 28 32 36

48 44 40 36 32 28 24 20 16 12

52 56 60 64 68 72 76 80 84 88

③ Answer these questions:

Nine times four is   36

Seven groups of four are   28

Four fives are   20

④ Dad took Devan, Jesse, and Owen to the fair. The roller coaster ride cost £1.50 for each person. How much did Dad have to pay for all of them to go on the ride?

£6

⑤ Work out these multiplication sums:

23	17	25	115	200	214
× 4	× 4	× 4	× 4	× 4	× 4
92	68	100	460	800	856

113

⑥ Divide each number by 4.

0 0   4 1   16 4   36 9   48 12

⑦ Jeff buys a pack of 4 pencils. The pack costs £1.68. How much does 1 pencil cost?   42p

⑧ Work out these division sums:

14
4⟌56

24
4⟌96

25
4⟌100

32
4⟌128

71
4⟌284

⑨ A box contains 24 chocolates. They are laid out in 4 equal rows. How many chocolates are there in each row?

6   chocolates

⑩ How many shapes are there in each group?

12   40

114

① How many sides do these shapes have in total?

50

24

48

56

② How many triangles have 27 angles in total?

9   triangles

③ The inside angles of an equilateral triangle add up to 180°. What is the value of each angle?

60°

④ Each side of a regular hexagon is 7 cm. What is the perimeter of the hexagon?

42 cm

⑤ What is the area of a rectangle with a length of 11 cm and a breadth of 4 cm?

44 cm²

⑥ Each angle of a square is 90°. What is the total of the 4 angles?

360°

115

⑦ A cuboid has 8 vertices. How many cuboids will have a total of 80 vertices?

10   cuboids

⑧ How many faces do these shapes have?

7 triangular prisms   35

9 cuboids   54

12 cylinders   36

⑨ How many edges do these shapes have?

7 cubes   84

4 square-based pyramids   32

3 hexagonal prisms   54

⑩ What is the volume of this cuboid?
Hint: Volume = length × breadth × height

72 cm³

6 cm
3 cm
4 cm

On these green pages, knowledge of times tables is applied to a general Maths concept. Shapes offer many opportunities for multiplying and dividing sums, such as working out perimeters, areas, and volumes.

Answers:

116–117 Groups of 5
118–119 Telling the time

116

1. A pack of greeting cards contains 5 cards. How many cards are there in 3 packs?
   15 cards

2. Complete each sequence:

   0 5 10 **15** **20** **25** **30** **35** **40** **45**

   60 55 50 **45** **40** **35** **30** **25** **20** **15**

   75 80 85 **90** **95** **100** **105** **110** **115** **120**

3. Answer these questions:
   Five groups of six are **30**
   Seven multiplied by five is **35**
   Eleven times five is **55**

4. David saved 24 5-pence coins. How much money did David save?
   £1.20

5. Work out these multiplication sums:

18	20	49	56	130	222
× 5	× 5	× 5	× 5	× 5	× 5
90	**100**	**245**	**280**	**650**	**1 110**

117

6. Divide each number by 5:
   10 **2**   25 **5**   30 **6**   50 **10**   85 **17**

7. Five children are given £1.95 to share equally among them. How much money will each child receive?
   39p

8. Work out these division sums:

   | **13** | **16** | **25** | **35** | **50** |
   | 5⟌65 | 5⟌80 | 5⟌125 | 5⟌175 | 5⟌250 |

9. There are 270 children in a school. There are 5 years, and each year has an equal number of children. How many children are there in Year 4?
   54 children

10. How many shapes are there in each group?
    25   **40**

118

1. How many minutes are there in 4 hours?
   240 minutes

2. How many minutes is it past 11 o'clock?
   45 minutes

3. How many minutes are there between 9.45 a.m. and 11.05 a.m.?
   80 minutes

4. How many minutes are there in 1 day?
   1 440 minutes

5. How many decades are there in half a century?
   Note: A decade is 10 years; a century is 100 years.
   5 decades

119

6. How many hours are there in these months?

   September (30 days) **720 hours**   February (28 days) **672 hours**

   May (31 days) **744 hours**

7. Write the number of minutes past the hour shown on each of these clocks.
   20 minutes   **55 minutes**   **10 minutes**   **35 minutes**

8. How many minutes are there between 3.10 p.m. and 5.25 p.m.?
   135 minutes

Telling the time and calculating the duration of time passing involves working with multiples of the five times table on occasion. The position of the minute hand can be quickly worked out by multiplying the numbers on a clock by five.

Answers:

122–123 Groups of 6
124–125 Bugs

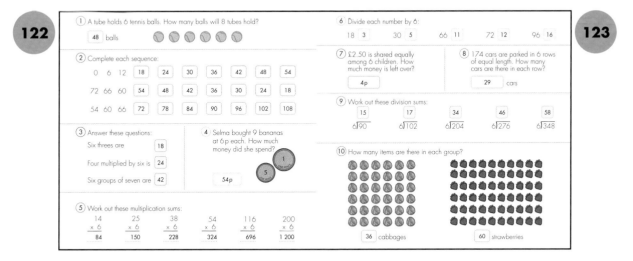

122

① A tube holds 6 tennis balls. How many balls will 8 tubes hold?

[48] balls

② Complete each sequence:

0   6   12   [18]   [24]   [30]   [36]   [42]   [48]   [54]

72   66   60   [54]   [48]   [42]   [36]   [30]   [24]   [18]

54   60   66   [72]   [78]   [84]   [90]   [96]   [102]   [108]

③ Answer these questions:

Six threes are [18]

Four multiplied by six is [24]

Six groups of seven are [42]

④ Selma bought 9 bananas at 6p each. How much money did she spend?

[54p]

⑤ Work out these multiplication sums:

14	25	38	54	116	200
× 6	× 6	× 6	× 6	× 6	× 6
84	150	228	324	696	1 200

123

⑥ Divide each number by 6:

18 [3]   30 [5]   66 [11]   72 [12]   96 [16]

⑦ £2.50 is shared equally among 6 children. How much money is left over?

[4p]

⑧ 174 cars are parked in 6 rows of equal length. How many cars are there in each row?

[29] cars

⑨ Work out these division sums:

6)90 = [15]   6)102 = [17]   6)204 = [34]   6)276 = [46]   6)348 = [58]

⑩ How many items are there in each group?

[36] cabbages   [60] strawberries

124

① Will recorded the number of bugs he saw in his garden in a month. How many legs did each type of insect have altogether?
Hint: An insect has 6 legs.

Name of bug	Number sighted	Total number of legs
Beetle	‖‖‖ ‖‖‖ ‖‖‖ ‖‖	102
Wasp	‖‖‖ ‖‖‖	48
Butterfly	‖‖‖ ‖‖‖ ‖‖‖	90
Ladybird	‖‖‖ ‖‖‖ ‖‖‖ ‖‖‖ ‖	126

② If a desert locust eats 2g of food each day, how much will a swarm of 66 million desert locusts eat in a day?

[132 million g]

③ A ladybird is about 6mm long. Under the microscope, the ladybird is magnified 40 times. What size is the ladybird when seen through the microscope?

[240mm or 24cm]

④ A leaf-cutter ant travels 360m each day. Altogether, what is the total distance the ant will travel in 60 days?

[21 600m or 21.6km]

125

⑤ How many wings does a swarm of 6 000 bees have altogether?
Hint: Bees have 4 wings.

[24 000] wings

⑥ A queen wasp can lay 2 000 eggs a day. How many eggs can she lay in 60 days?

[120 000] eggs

⑦ A butterfly lays 600 eggs, and only a quarter of them hatch into caterpillars. How many eggs **do not** hatch?

[450] eggs

⑧ Class 5F went to a pond. They made a picture chart of the number of bugs they saw in the pond. How many bugs of each type did they see?

⬤ = 6 bugs   ◗ = 3 bugs

Name of bug	Number of bugs	Total
Pond skaters	⬤ ⬤ ⬤ ⬤	24
Water bugs	⬤ ⬤ ⬤	18
Whirligig beetles	⬤ ⬤ ◗	15
Dragonfly nymphs	◗	3
Water spiders	⬤	6

Children will be familiar with a range of tables and charts to represent data and use them to interpret and calculate information. The tables on these pages show a tally chart, using the "five bar gate" method of recording amounts, and a pictogram that uses an image to represent the data collected.

Answers:

126–127 Sports
128–129 Groups of 7

126

1. Adam ran 400 m in 59 seconds. Jonas took twice as long. How long did Jonas take?

   `118 seconds`

2. Three cyclists raced at 58 mph, 63 mph, and 56 mph. What was their average speed?

   `59 mph`

3. These were the results of a season's football games: A win = 5 points, a draw = 3 points, and a loss = 1 point. How many points did each team get?

Football team	Win	Loss	Draw	Points
England	8	2	5	57
United States	7	3	6	56
Japan	7	1	7	57
Brazil	9	2	4	59

4. The winner of a tennis tournament won 4 times the prize money of a semi-finalist. If a semi-finalist received £475 000, how much money did the winner receive?

   £1 900 000

127

5. John threw a javelin a distance of 64 m. Amy threw the javelin an eighth ($\frac{1}{8}$) less than John's distance. How far did Amy throw?

   `56 m`

6. Seven race car drivers have a total of 1 645 points. What is the average number of points scored?

   `235` points

7. There were 162 rounders players taking part in a tournament. Each team had 9 players. How many teams were there?

   `18` teams

8. The length of a swimming pool is 25 m. This chart shows how many times each child swam that length. How far did each child swim?

Name of child	Number of lengths	Total distance
Harry	10	250 m
Jasmine	8	200 m
Jamie	6	150 m
Heidi	5	125 m

These yellow pages are also themed and involve using a mix of times tables to solve the sums. Question 2 involves knowing ways to calculate averages. The amounts need to be added and then divided by the number of amounts, which is three in this question.

128

1. A dog eats 3 dog treats a day. How many treats will it eat in 7 days?

   `21` treats

2. Complete each sequence:

   0   7   14   `21`   `28`   `35`   `42`   `49`   `56`   `63`

   84   77   70   `63`   `56`   `49`   `42`   `35`   `28`   `21`

   35   42   49   `56`   `63`   `70`   `77`   `84`   `91`   `98`

3. Answer these questions:

   Seven sixes are `42`

   Eight multiplied by seven is `56`

   Five groups of seven are `35`

4. A train ticket costs £7. How much will 6 tickets cost?

   £42

5. Work out these multiplication sums:

   | 14 | 20 | 35 | 59 | 123 | 246 |
× 7	× 7	× 7	× 7	× 7	× 7
98	140	245	413	861	1 722

129

6. Divide each number by 7:

   0 `0`   21 `3`   49 `7`   77 `11`   98 `14`

7. Seven books cost £35.84 altogether. If each book was the same price, what was the price of 1 book?

   £5.12

8. Work out these division sums:

   `12`   `20`   `15`   `19`   `32`
   7)84   7)140   7)105   7)133   7)224

9. Share 42 chairs equally around 7 tables. How many chairs will you keep around each table?

   `6` chairs

10. How many shapes are there in each group?

    `49`   `21`

Some of the time fillers provide tips on how to calculate or check the answers to certain times tables. Knowing a variety of methods supports understanding of the concepts of multiplication and division.

Answers:

130–131 Days of the week
132–133 Dice and cards

130

1 How many days are there in 15 weeks?

`105` days

2 How many weeks are there in 7 years?

`364` weeks

3 How many hours are there in a week?

`168 hours`

4 Dad works 35 hours a week. How many hours does he work over 4 weeks?

`140 hours`

5 Fran cycles for 30 minutes every day. How many minutes does she cycle in one week?

`210 minutes`

6 A bookshop opens for 7 hours each day from Monday to Saturday. How many hours is it open in one week?

`42 hours`

131

7 Chris practises on the keyboard for 105 minutes every week. He does an equal amount of time every day. How long is each practice?

`15 minutes`

8 Ella travelled for 91 days. How many weeks is this?

`13` weeks

9 Dad books a holiday 22 weeks before going. How many days does the family have to wait?

`154` days

10 How long in minutes does Kim do these daily activities in one week?

Watching 45 minutes of television — 315 minutes

Playing 30 minutes of computer games — 210 minutes

Reading for 1 hour 10 minutes — 490 minutes

132

1 Multiply the two numbers shown on the dice:

× = `24`   × = `25`

× = `12`   × = `6`

2 Add the numbers shown on the dice, and then multiply your answer by 6.

{ + } × 6 = `30`   { + } × 6 = `42`

{ + } × 6 = `72`   { + } × 6 = `48`

3 Jack threw a double six 5 times. What was his total?

`60`

4 These are the scores of four players. They each need to throw a double to reach 100 points. What is the number that needs to appear on both dice for each player? Fill in the spaces in the table.

Player	Score	Number required on both dice
1	90	5
2	98	1
3	94	3
4	88	6

133

5 Jess throws a die 100 times and records her scores. What is the total amount scored altogether by Jess? Fill in the spaces in the table.

Number on die	Number of times thrown	Total
1	༔ ༔ ༔ I	16
2	༔ ༔ ༔ ༔	40
3	༔ ༔ ༔ III	54
4	༔ ༔ ༔	60
5	༔ ༔ ༔ II	85
6	༔ ༔ IIII	84
	Total	339

6 A full deck of cards has 13 cards of each suit.
Note: There are 4 suits in a deck.
How many cards are there in a deck? `52` cards

How many cards are there in

4 decks? `208` cards

9 decks? `468` cards

6 decks? `312` cards

Multiply each of these cards by 8:

9 hearts `72`

8 diamonds `64`

Queen (12) clubs `96`

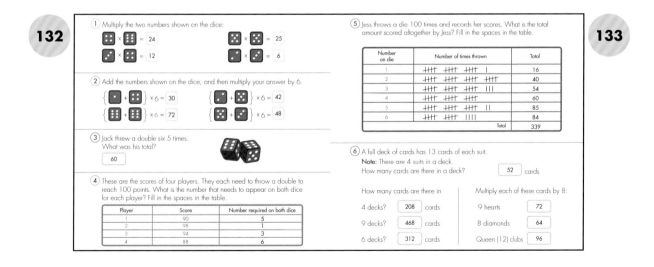

134–135 Groups of 8
136–137 Solar system

134

① Six trains run every hour. Each train pulls 8 carriages. How many carriages are pulled every hour?

48 carriages

② Complete these sequences:

0 8 16 | 24 | 32 | 40 | 48 | 56 | 64 | 72 |

96 88 80 | 72 | 64 | 56 | 48 | 40 | 32 | 24 |

40 48 56 | 64 | 72 | 80 | 88 | 96 | 104 | 112 |

③ Answer these questions:

Eight sixes are 48

Two multiplied by eight is 16

Nine times eight is 72

④ A bag of apples costs £1.46. How much will 8 bags cost?

£11.68

⑤ Work out these multiplication sums:

15	24	48	97	120	236
× 8	× 8	× 8	× 8	× 8	× 8
120	192	384	776	960	1 888

135

⑥ Work out these division sums:

| 12 | 18 | 21 | 32 | 39 |
| 8)96 | 8)144 | 8)168 | 8)256 | 8)312 |

⑦ Tammy needs 192 m of fencing to go around her garden. Each fencing panel is 8 m long. How many panels will she need?

24 panels

⑧ Perry pays £2.80 for 8 pencils. How much did 1 pencil cost?

35p

⑨ Divide each number by 8:

0 0   24 3   56 7   80 10   96 12

⑩ How many gems are there in each group?

64

16

136

① Thirty-five satellites are launched into orbit each year. How many satellites have been launched in 8 years?

280 satellites

② A probe travelling at 500 mph will take 8 years to get to Mars. How fast does the probe need to travel to get there in 1 year?

4 000 mph

③ Mars takes approximately 687 days to orbit the Sun. How many days will it take for Mars to orbit the Sun 8 times?

5 496 days

④ Neptune takes 165 years to orbit the Sun. How many years will it take to go around the Sun 8 times?

1 320 years

⑤ Mercury takes 88 Earth days to orbit the Sun. How many times will Mercury go around the Sun in 880 Earth days?

10 times

137

⑥ Multiply the 8 moons of Neptune with the 16 moons of Jupiter.

128 moons

⑦ The midday surface temperature of Mercury is 420°C. What is 8 times hotter than the surface of Mercury?

3 360°C

⑧ Mercury is 4 878 km in diameter. What is an eighth ($\frac{1}{8}$) of its diameter?

609.75 km

⑨ One day on Saturn lasts 10 hours and 14 minutes on Earth. How long in Earth time will 8 days on Saturn last?

81 hours 52 minutes

One day on Uranus lasts 17 hours and 8 minutes on Earth. How long in Earth time will 8 days on Saturn last?

137 hours 4 minutes

⑩ Saturn is 1.4 billion km away from the Sun. What is $\frac{5}{8}$ of this distance?

875 million km

Jupiter is 780 million km away from the Sun. What is $\frac{3}{8}$ of this distance?

292.5 million km

Answers:

138–139 Fractions
140–141 Groups of 9

138

1. What is half ($\frac{1}{2}$) of each number?

   18 **9**   10 **5**   6 **3**   24 **12**

2. What is a third ($\frac{1}{3}$) of each amount?

   12g **4g**   27g **9g**   33g **11g**   42g **14g**

3. What is a quarter ($\frac{1}{4}$) of each number?

   4 **1**   20 **5**   36 **9**   52 **13**

4. There are 60 carrots in a box. How many carrots make up…

   $\frac{7}{10}$ of the box? **42** carrots

   $\frac{1}{10}$ of the box? **6** carrots $\frac{2}{10}$ of the box? **12** carrots

5. There were 25 bananas, and $\frac{1}{5}$ were eaten. How many bananas are left?

   20 bananas

139

6. What is three quarters ($\frac{3}{4}$) of each number?

   12 **9**   24 **18**   32 **24**   44 **33**

7. What is $\frac{1}{8}$ of 48 slices of pizza?

   6 slices

8. What is $\frac{7}{10}$ of 40?

   28

9. There are 30 children in a class. $\frac{3}{5}$ of the class have school dinners. How many children **do not** have school dinners?

   12 children

10. Oliver picked 54 apples. $\frac{1}{6}$ were rotten. How many apples were rotten?

    9 apples

At this level, children should be familiar with calculating fractions of amounts. Make sure they read the question carefully so that they don't get tricked by what the question is asking. For example, Question 9 asks for the number of children who do not have school lunches, so they need to calculate $\frac{2}{5}$ of the 30 children.

140

1. There are 8 horse races in a day. If 9 different horses took part in each race, how many horses ran that day?

   72 horses

2. Complete each sequence:

   0 9 18 **27** **36** **45** **54** **63** **72** **81**

   108 99 90 **81** **72** **63** **54** **45** **36** **27**

   45 54 63 **72** **81** **90** **99** **108** **117** **126**

3. Answer these questions:

   Three multiplied by nine is **27**

   Nine eights are **72**

   Six groups of nine are **54**

4. A bunch of flowers costs £4.99. How much will 9 bunches cost?

   £44.91

5. Work out these multiplication sums:

16	23	92	47	150	218
× 9	× 9	× 9	× 9	× 9	× 9
144	**207**	**828**	**423**	**1 350**	**1 962**

141

6. Divide each number by 9:

   9 **1**   36 **4**   45 **5**   90 **10**   108 **12**

7. Jake needed £468 to buy a new television. He decided to save an equal amount over 9 weeks to reach the total. What is the amount he needed to save each week?

   £52

8. How many shapes are there in each group?

   63   **18**

Answers:

142–143 Shopping
146–147 Division

142

① Calculate the total cost that Karl spent shopping.

Item	Cost per item	Amount	Total
Tomatoes	20p	6	£1.20
Carrots	10p	8	£0.80
Cabbage	89p	2	£1.78
Peppers	43p	5	£2.15
Cheese	£1.26	1	£1.26
Bread	76p	3	£2.28
Juice	£1.49	4	£5.96
Milk	72p	6	£4.32
Biscuits	89p	7	£6.23
Pasta	£2.56	2	£5.12
			£31.10

143

② A shopkeeper sold 6 red coats at £89 each. How much did the red coats cost altogether?

£534

③ Mum bought 3 bracelets at £7.84 each. How much money did Mum spend?

£23.52

④ Tami spent £77.97 on 3 pairs of shoes. Each pair cost the same amount. How much did each pair cost?

£25.99

⑤ In a sale, the cost of a hat was reduced by 20%. The original price of the hat was £14.50. How much was it reduced by?

£2.90

These pages demonstrate how multiplication and division skills are useful in shopping situations. Knowing how to calculate price reductions in sales can be very helpful when deciding which products to buy.

146

① Match each question to its answer:

168 ÷ 6 → 28
524 ÷ 4 → 131
595 ÷ 7 → 85
729 ÷ 9 → 81

② Use the long division method to work out each sum:

4)648	2)496	5)760	6)822
162	248	152	137
−4	−4	−5	−6
24	09	26	22
−24	−08	−25	−18
08	16	10	42
−08	−16	−10	−42
0	0	0	0

③ What is the remainder each time?

592 ÷ 3 = 1
264 ÷ 7 = 5
786 ÷ 4 = 2
543 ÷ 9 = 3

④ Circle all the multiples of 7:

(14) 23 (35) 43 76 (84)

147

⑤ Circle all the multiples of 9:

28 (54) 61 83 (99) (108)

⑥ Circle all the multiples of 12:

(24) 45 (60) 56 (72) 98 (132)

⑦ Work out these money sums:

£14.58 ÷ 3 = £4.86
£35.60 ÷ 8 = £4.45
£26.96 ÷ 4 = £6.74
£66.69 ÷ 9 = £7.41

⑧ List all the factors for each number:

24: 1, 2, 3, 4, 6, 8, 12, 24
36: 1, 2, 3, 4, 6, 9, 12, 18, 36
72: 1, 2, 3, 4, 6, 8, 9, 12, 18, 24, 36, 72
100: 1, 2, 4, 5, 10, 20, 25, 50, 100

Children may know either one method or a range of methods to easily divide multi-digit whole numbers, such as the long division method. Also, they should be able to recognise factor pairs and the whole numbers as multiples of each of its factors.

Answers:

148–149 Groups of 11
150–151 Sequences

① A farmer plants 6 rows of tulips, with 11 bulbs in each row. How many tulip bulbs are planted?

[66] bulbs

② Complete each sequence:

0　11　22　[33]　[44]　[55]　[66]　[77]　[88]　[99]

143　132　121　[110]　[99]　[88]　[77]　[66]　[55]　[44]

66　77　88　[99]　[110]　[121]　[132]　[143]　[154]　[165]

③ Answer these questions:

Eleven fours are [44]

Eleven groups of seven are [77]

Twelve times eleven is [132]

④ Ellie buys 11 T-shirts at £1.10 each. How much does Ellie pay for 11 T-shirts?

£12.10

⑤ Work out these multiplication sums:

14	25	69	33	81	100
×11	×11	×11	×11	×11	×11
154	275	759	363	891	1 100

⑥ Divide each number by 11:

22　2　　88　8　　121　11　　143　13　　176　16

⑦ Work out these division sums:

[17]　　[27]　　[33]　　[52]　　[71]

11)187　　11)297　　11)363　　11)572　　11)781

⑧ How many spots are there in each group?

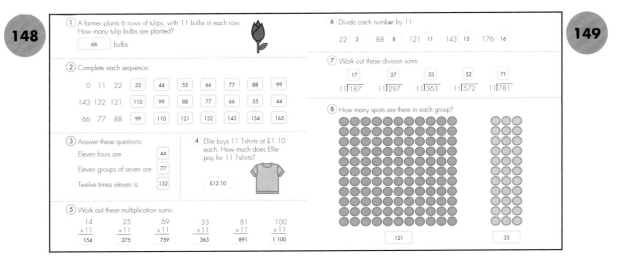

[121]　　　　[33]

① Fill in the missing numbers in each sequence:

5　10　[15]　[20]　[25]　30　35　[40]　45　[50]

60　[65]　[70]　75　[80]　[85]　90　95　[100]　105

② Complete each sequence:

4　8　12　[16]　[20]　[24]　[28]　[32]　[36]　[40]

7　14　21　[28]　[35]　[42]　[49]　[56]　[63]　[70]

25　50　75　[100]　[125]　[150]　[175]　[200]　[225]　[250]

③ Continue this pattern:

④ Complete this chart:

×	0	1	2	3	4	5	6	7	8	9	10
6	0	6	12	18	24	30	36	42	48	54	60
9	0	9	18	27	36	45	54	63	72	81	90

⑤ Fill in the missing numbers in each sequence:

80　72　[64]　[56]　[48]　40　[32]　24　[16]　8

60　56　52　[48]　[44]　40　36　[32]　[28]　24

⑥ Continue this pattern:

⑦ Complete this chart:

×	10	9	8	7	6	5	4	3	2	1	0
11	110	99	88	77	66	55	44	33	22	11	0
12	120	108	96	84	72	60	48	36	24	12	0

⑧ Complete each sequence:

200　190　180　[170]　[160]　[150]　[140]　[130]　[120]　[110]

150　148　146　[144]　[142]　[140]　[138]　[136]　[134]　[132]

These sequences identify multiples of times tables and also the patterns between numbers.

Question 6 draws on knowledge of square numbers where a number is multiplied by itself, such as 5 × 5 = 25.

170 Answers:

152–153 Groups of 12
154–155 Dozen a day

152

1. A baker takes an hour to cook 12 loaves of bread. How many loaves can he make in 3 hours?

   `36` loaves

2. Complete each sequence:

   0 12 24 `36` `48` 60 72 84 96 `108`

   144 132 120 `108` `96` 84 72 60 48 `36`

   60 72 84 `96` `108` `120` `132` `144` `156` `168`

3. Answer these questions:

   Five groups of twelve are `60`

   Eight times twelve is `96`

   Ten multiplied by twelve is `120`

4. Cara collects trading cards. She buys 12 packs of 4 cards at £2 per pack. How much money does Cara spend and how many trading cards will she have?

   `£24` `48` trading cards

5. Work out these multiplication sums.

13	17	24	35	42	100
×12	×12	×12	×12	×12	×12
156	204	288	420	504	1 200

153

6. Divide each number by 12:

   0 `0`   48 `4`   84 `7`   132 `11`   192 `16`

7. Mum has a loan of £864, which she pays back in equal amounts over 12 months. How much does Mum pay each month?

   `£72.00`

8. How many shapes are there in each group?

   `144`

   `60`

154

1. A dozen children split themselves equally into 3 teams to play a game. How many children are there in each team?

   `4` children

2. Cupcakes were sold in boxes of 12. How many cupcakes were there in 15 boxes?

   `180` cupcakes

3. How many dozen eggs are there in a gross of eggs? **Hint:** A gross is 144.

   `12 dozen`

4. What are the factors of 12?

   1   2   3   4   6   12

5. A score of children had 12 sweets each. How many sweets did they have altogether? **Note:** A score is 20.

   `240` sweets

155

6. A chef cooks a batch of 12 pancakes in 12 minutes. How many batches of 12 pancakes can he make in an hour?

   `5` batches

7. Trains arrived at Whistlestop Station 3 times an hour. How many trains arrived in 12 hours?

   `36` trains

8. A group of musicians performed a dozen pieces in a concert. Each piece lasted 4 minutes. How many minutes did the musicians perform altogether?

   `48 minutes`

9. 900 raffle tickets were sold at a fund-raising event. There were a dozen prizes. What was the chance of winning a prize? Circle the correct answer.

   1 in 50   (1 in 75)   1 in 100

10. Once a month, Jill ran a distance of 10 km in a cross-country event. How many kilometres did Jill run in a year?

    `120 km`

Answers:

120–121 Beat the clock 1
144–145 Beat the clock 2

These Beat the clock pages test your child's ability to quickly recall times tables facts. The tests require your child to work under some pressure. As with most tests of this type, tell children before they start not to get stuck on one question, but to move on and return to the tricky one later if time allows. Encourage your child to record his/her score and the time taken to complete the test, then to retake the test later to see if he/she can improve on his/her previous attempt.

120 / 121

(1) 0 x 3 = 0	(2) 12 x 2 = 24	(3) 9 x 5 = 45	(31) 24 ÷ 3 = 8	(32) 6 ÷ 2 = 3	(33) 15 ÷ 5 = 3
(4) 1 x 4 = 4	(5) 11 x 5 = 55	(6) 3 x 4 = 12	(34) 30 ÷ 5 = 6	(35) 9 ÷ 3 = 3	(36) 30 ÷ 10 = 3
(7) 7 x 3 = 21	(8) 10 x 2 = 20	(9) 2 x 2 = 4	(37) 14 ÷ 2 = 7	(38) 3 ÷ 3 = 1	(39) 50 ÷ 10 = 5
(10) 6 x 5 = 30	(11) 11 x 4 = 44	(12) 8 x 2 = 16	(40) 40 ÷ 5 = 8	(41) 2 ÷ 2 = 1	(42) 90 ÷ 10 = 9
(13) 4 x 4 = 16	(14) 10 x 5 = 50	(15) 2 x 5 = 10	(43) 21 ÷ 3 = 7	(44) 6 ÷ 3 = 2	(45) 10 ÷ 10 = 1
(16) 6 x 3 = 18	(17) 10 x 9 = 90	(18) 9 x 4 = 36	(46) 18 ÷ 2 = 9	(47) 27 ÷ 3 = 9	(48) 80 ÷ 10 = 8
(19) 1 x 5 = 5	(20) 12 x 4 = 48	(21) 6 x 4 = 24	(49) 15 ÷ 5 = 3	(50) 15 ÷ 3 = 5	(51) 60 ÷ 10 = 6
(22) 7 x 4 = 28	(23) 10 x 1 = 10	(24) 12 x 10 = 120	(52) 40 ÷ 8 = 5	(53) 36 ÷ 3 = 12	(54) 70 ÷ 10 = 7
(25) 2 x 4 = 8	(26) 10 x 7 = 70	(27) 10 x 10 = 100	(55) 18 ÷ 3 = 6	(56) 60 ÷ 5 = 12	(57) 110 ÷ 10 = 11
(28) 8 x 4 = 32	(29) 11 x 2 = 22	(30) 0 x 2 = 0	(58) 30 ÷ 3 = 10	(59) 25 ÷ 5 = 5	(60) 100 ÷ 10 = 10

144 / 145

(1) 0 x 9 = 0	(2) 0 x 6 = 0	(3) 8 x 9 = 72	(31) 0 ÷ 7 = 0	(32) 60 ÷ 6 = 10	(33) 63 ÷ 9 = 7
(4) 5 x 8 = 40	(5) 7 x 8 = 56	(6) 3 x 8 = 24	(34) 0 ÷ 8 = 0	(35) 99 ÷ 9 = 11	(36) 84 ÷ 7 = 12
(7) 2 x 9 = 18	(8) 7 x 6 = 42	(9) 2 x 7 = 14	(37) 9 ÷ 9 = 1	(38) 27 ÷ 9 = 3	(39) 48 ÷ 8 = 6
(10) 1 x 7 = 7	(11) 9 x 9 = 81	(12) 10 x 9 = 90	(40) 6 ÷ 6 = 1	(41) 96 ÷ 8 = 12	(42) 72 ÷ 8 = 9
(13) 2 x 8 = 16	(14) 1 x 8 = 8	(15) 12 x 6 = 72	(43) 21 ÷ 7 = 3	(44) 32 ÷ 8 = 4	(45) 63 ÷ 7 = 9
(16) 3 x 6 = 18	(17) 9 x 7 = 63	(18) 10 x 7 = 70	(46) 24 ÷ 6 = 4	(47) 45 ÷ 9 = 5	(48) 35 ÷ 7 = 5
(19) 4 x 7 = 28	(20) 4 x 9 = 36	(21) 11 x 6 = 66	(49) 77 ÷ 7 = 11	(50) 48 ÷ 6 = 8	(51) 54 ÷ 9 = 6
(22) 8 x 8 = 64	(23) 6 x 8 = 48	(24) 11 x 8 = 88	(52) 80 ÷ 8 = 10	(53) 49 ÷ 7 = 7	(54) 12 ÷ 6 = 2
(25) 9 x 6 = 54	(26) 6 x 7 = 42	(27) 12 x 7 = 84	(55) 36 ÷ 6 = 6	(56) 64 ÷ 8 = 8	(57) 56 ÷ 7 = 8
(28) 5 x 6 = 30	(29) 7 x 7 = 49	(30) 12 x 9 = 108	(58) 40 ÷ 8 = 5	(59) 81 ÷ 9 = 9	(60) 108 ÷ 9 = 12

156–157 Beat the clock 3

(1) 3 × 9 = 27	(2) 19 × 4 = 76	(3) 20 × 6 = 120	(31) 54 ÷ 9 = 6	(32) 245 ÷ 5 = 49	(33) 85 ÷ 5 = 17
(4) 1 × 7 = 7	(5) 12 × 9 = 108	(6) 12 × 8 = 96	(34) 81 ÷ 9 = 9	(35) 112 ÷ 8 = 14	(36) 24 ÷ 3 = 8
(7) 4 × 6 = 24	(8) 10 × 8 = 80	(9) 12 × 5 = 60	(37) 56 ÷ 8 = 7	(38) 100 ÷ 5 = 20	(39) 92 ÷ 2 = 46
(10) 5 × 4 = 20	(11) 11 × 0 = 0	(12) 11 × 2 = 22	(40) 56 ÷ 7 = 8	(41) 108 ÷ 6 = 18	(42) 63 ÷ 3 = 21
(13) 5 × 8 = 40	(14) 11 × 6 = 66	(15) 18 × 3 = 54	(43) 42 ÷ 6 = 7	(44) 456 ÷ 1 = 456	(45) 28 ÷ 2 = 14
(16) 8 × 2 = 16	(17) 12 × 4 = 48	(18) 16 × 2 = 32	(46) 91 ÷ 7 = 13	(47) 537 ÷ 3 = 179	(48) 76 ÷ 2 = 38
(19) 7 × 7 = 49	(20) 10 × 5 = 50	(21) 14 × 0 = 0	(49) 40 ÷ 4 = 10	(50) 860 ÷ 10 = 86	(51) 99 ÷ 11 = 9
(22) 7 × 3 = 21	(23) 15 × 9 = 135	(24) 32 × 1 = 32	(52) 25 ÷ 5 = 5	(53) 240 ÷ 12 = 20	(54) 320 ÷ 10 = 32
(25) 1 × 3 = 3	(26) 17 × 8 = 136	(27) 19 × 10 = 190	(55) 51 ÷ 3 = 17	(56) 143 ÷ 11 = 13	(57) 144 ÷ 12 = 12
(28) 2 × 10 = 20	(29) 15 × 7 = 105	(30) 16 × 12 = 192	(58) 64 ÷ 4 = 16	(59) 121 ÷ 11 = 11	(60) 1000 ÷ 10 = 100

Times Tables

PRACTICE

Author Sue Phillips
Consultant Sean McArdle

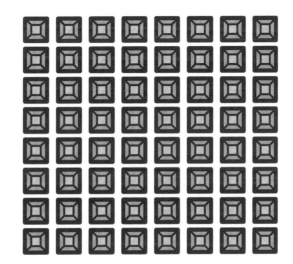

Contents

2x table

Count in 2s, colour, and find a pattern.

1	2	3	4	5
6	7	8	9	10
11	12	13	14	15
16	17	18	19	20
21	22	23	24	25

Write the answers.

1 x 2 = 2   2 x 2 = ☐   3 x 2 = ☐   4 x 2 = ☐

5 x 2 = ☐   6 x 2 = ☐   7 x 2 = ☐   8 x 2 = ☐

9 x 2 = ☐   10 x 2 = ☐   11 x 2 = ☐   12 x 2 = ☐

How many ears?

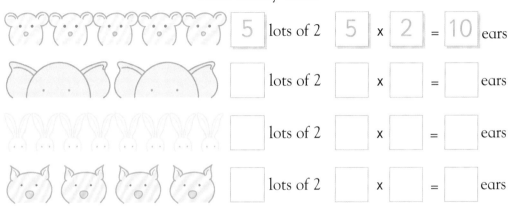

5 lots of 2   5 x 2 = 10 ears

☐ lots of 2   ☐ x ☐ = ☐ ears

☐ lots of 2   ☐ x ☐ = ☐ ears

☐ lots of 2   ☐ x ☐ = ☐ ears

Multiplying by 2

Write the sums.

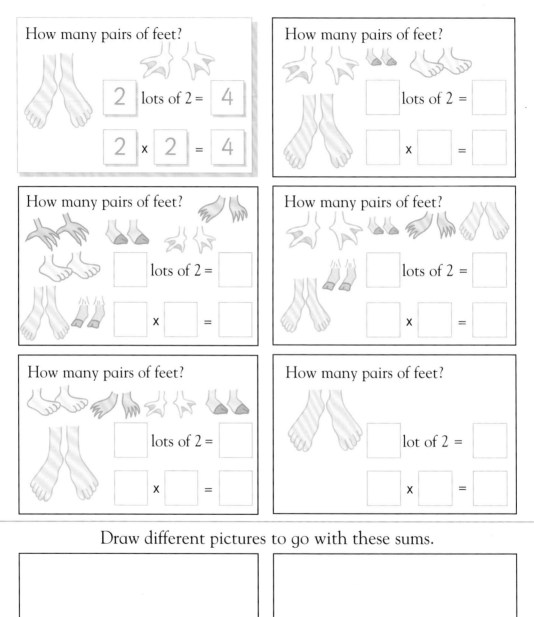

How many pairs of feet?

| 2 | lots of 2 = | 4 |

| 2 | x | 2 | = | 4 |

How many pairs of feet?

| | lots of 2 = | |

| | x | | = | |

How many pairs of feet?

| | lots of 2 = | |

| | x | | = | |

How many pairs of feet?

| | lots of 2 = | |

| | x | | = | |

How many pairs of feet?

| | lots of 2 = | |

| | x | | = | |

How many pairs of feet?

| | lot of 2 = | |

| | x | | = | |

Draw different pictures to go with these sums.

8 x 2 = 16

10 x 2 = 20

Dividing by 2

Share the eggs equally between the nests.

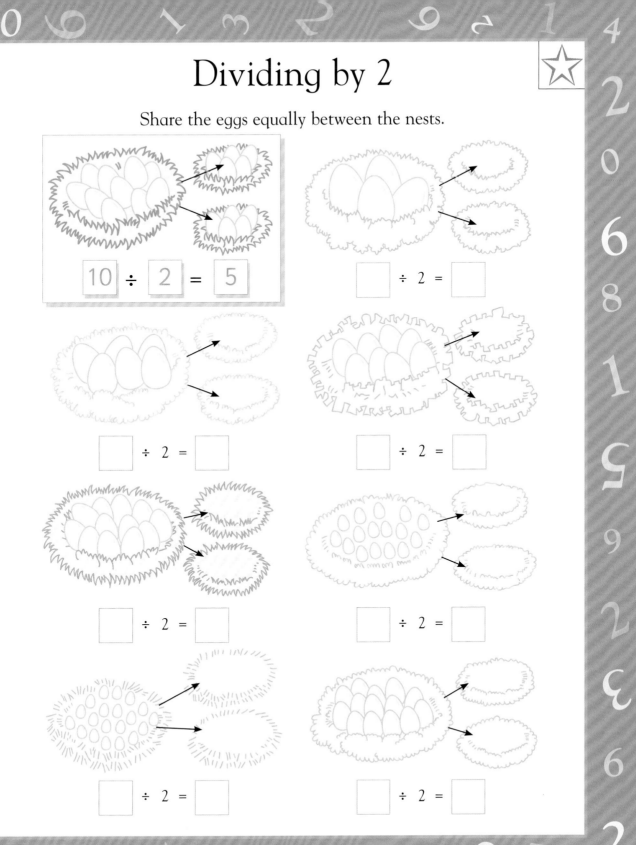

$10 \div 2 = 5$

$\square \div 2 = \square$

$\square \div 2 = \square$

$\square \div 2 = \square$

$\square \div 2 = \square$

$\square \div 2 = \square$

$\square \div 2 = \square$

$\square \div 2 = \square$

Using the 2x table

Write the sums to match the stamps.

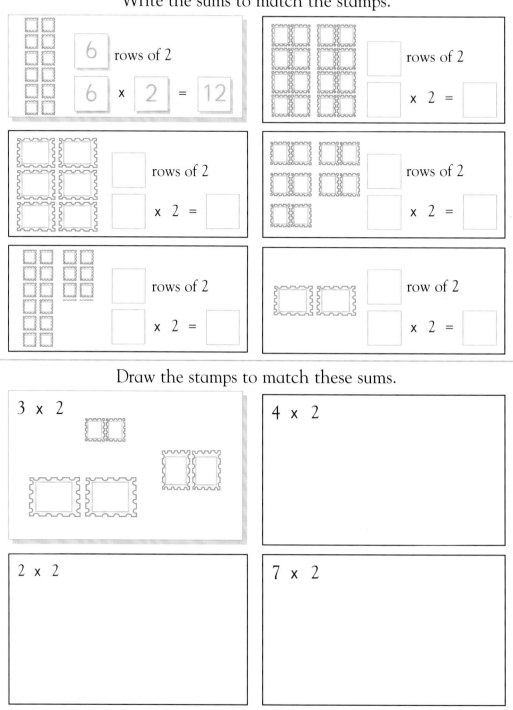

6 rows of 2

6 x 2 = 12

[ ] rows of 2

[ ] x 2 = [ ]

[ ] rows of 2

[ ] x 2 = [ ]

[ ] rows of 2

[ ] x 2 = [ ]

[ ] rows of 2

[ ] x 2 = [ ]

[ ] row of 2

[ ] x 2 = [ ]

Draw the stamps to match these sums.

3 x 2

4 x 2

2 x 2

7 x 2

Using the 2x table

Each face stands for 2. Join each set of faces to the correct number.

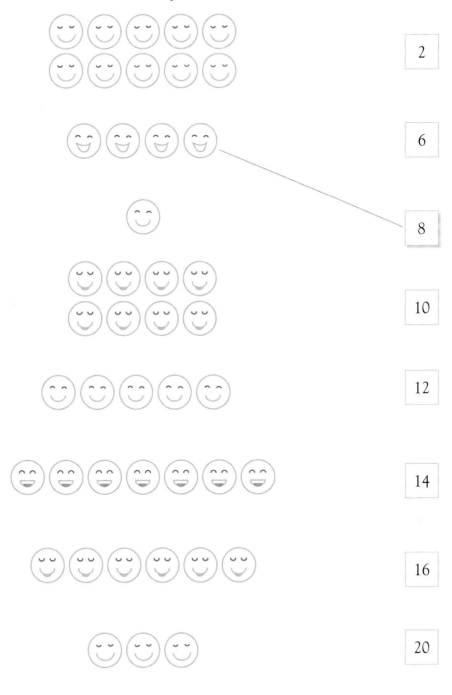

2

6

8

10

12

14

16

20

Using the 2x table

How many eyes?

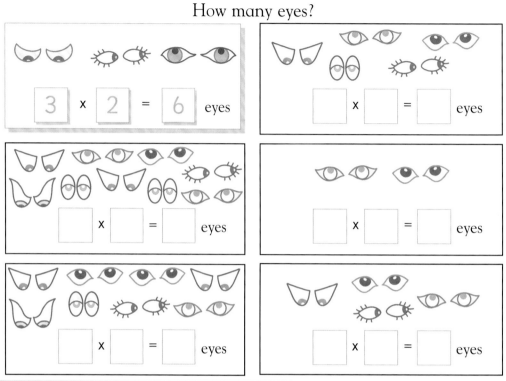

| 3 | x | 2 | = | 6 | eyes |

| | x | | = | | eyes |

| | x | | = | | eyes |

| | x | | = | | eyes |

| | x | | = | | eyes |

| | x | | = | | eyes |

Draw your own pictures to match these number sentences.

2 x 2 = 4

10 x 2 = 20

3 x 2 = 6

7 x 2 = 14

5x table

Count in 5s, colour, and find a pattern.

1	2	3	4	5	6	7	8	9	10
11	12	13	14	15	16	17	18	19	20
21	22	23	24	25	26	27	28	29	30
31	32	33	34	35	36	37	38	39	40
41	42	43	44	45	46	47	48	49	50
51	52	53	54	55	56	57	58	59	60
61	62	63	64	65	66	67	68	69	70
71	72	73	74	75	76	77	78	79	80
81	82	83	84	85	86	87	88	89	90
91	92	93	94	95	96	97	98	99	100

Write the answers.

1 x 5 = 5   2 x 5 = ☐   3 x 5 = ☐   4 x 5 = ☐

5 x 5 = ☐   6 x 5 = ☐   7 x 5 = ☐   8 x 5 = ☐

9 x 5 = ☐   10 x 5 = ☐   11 x 5 = ☐   12 x 5 = ☐

How many sweets?

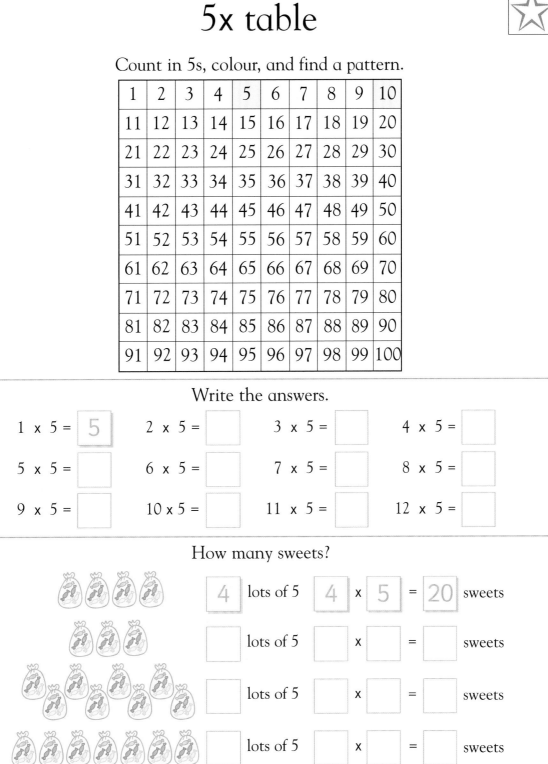

4 lots of 5   4 x 5 = 20 sweets

☐ lots of 5   ☐ x ☐ = ☐ sweets

☐ lots of 5   ☐ x ☐ = ☐ sweets

☐ lots of 5   ☐ x ☐ = ☐ sweets

Multiplying by 5

Draw a ring around rows of 5. Complete the sum.

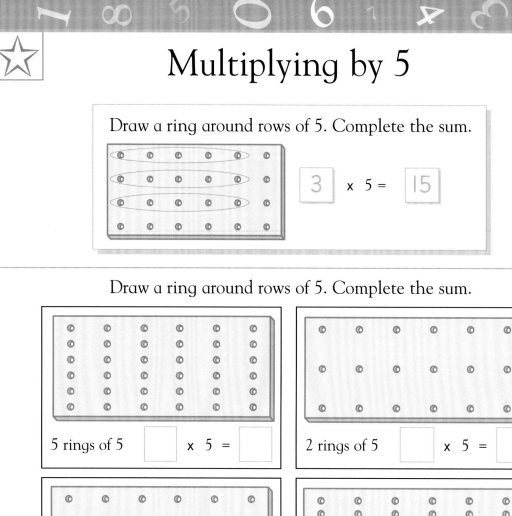

3 x 5 = 15

Draw a ring around rows of 5. Complete the sum.

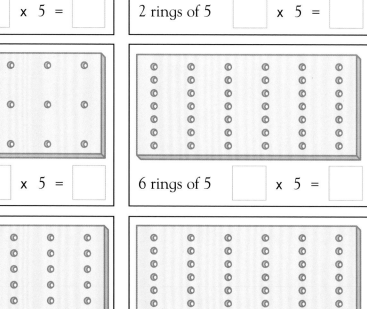

5 rings of 5 ☐ x 5 = ☐

2 rings of 5 ☐ x 5 = ☐

1 ring of 5 ☐ x 5 = ☐

6 rings of 5 ☐ x 5 = ☐

4 rings of 5 ☐ x 5 = ☐

7 rings of 5 ☐ x 5 = ☐

Dividing by 5

Write a number sentence to show how many cubes are in each tower.

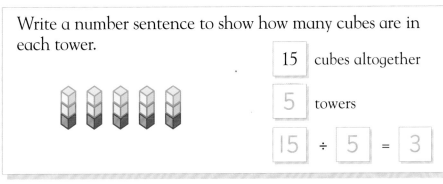

15 cubes altogether

5 towers

15 ÷ 5 = 3

Write a number sentence to show how many cubes are in each tower.

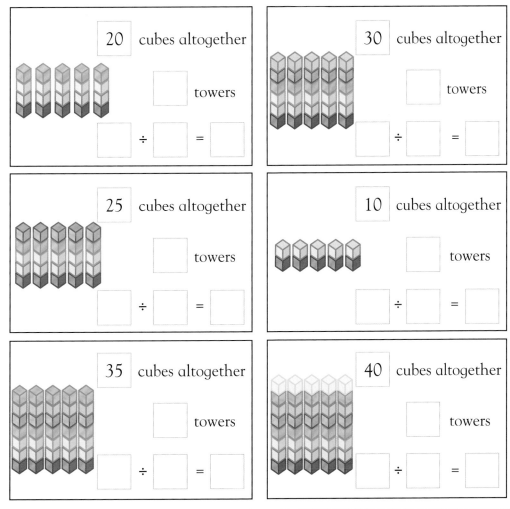

20 cubes altogether

towers

☐ ÷ ☐ = ☐

30 cubes altogether

towers

☐ ÷ ☐ = ☐

25 cubes altogether

towers

☐ ÷ ☐ = ☐

10 cubes altogether

towers

☐ ÷ ☐ = ☐

35 cubes altogether

towers

☐ ÷ ☐ = ☐

40 cubes altogether

towers

☐ ÷ ☐ = ☐

Using the 5x table

Write the number that is hiding under the star.

 x 5 = 10

3 x 5 =

 x 5 = 25

1 x 5 =

 x 5 = 50

8 x 5 =

 x 5 = 45

0 x 5 =

x 5 = 35

6 x 5 =

Using the 5x table

Each frog stands for 5. Join each set of frogs to the correct number.

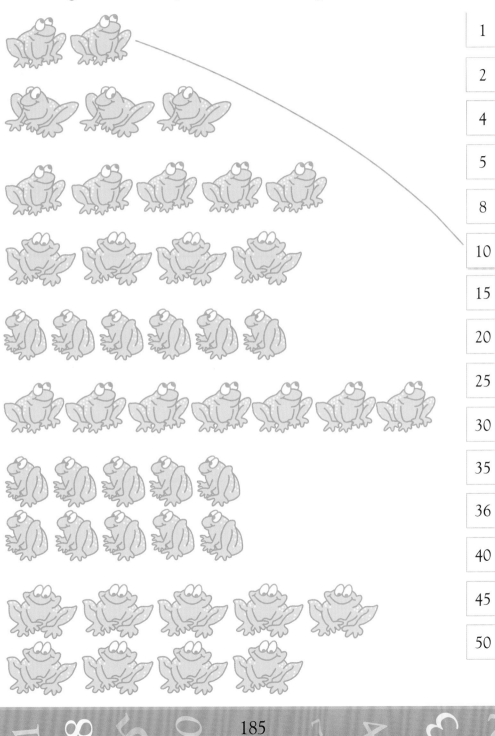

1

2

4

5

8

10

15

20

25

30

35

36

40

45

50

Using the 5x table

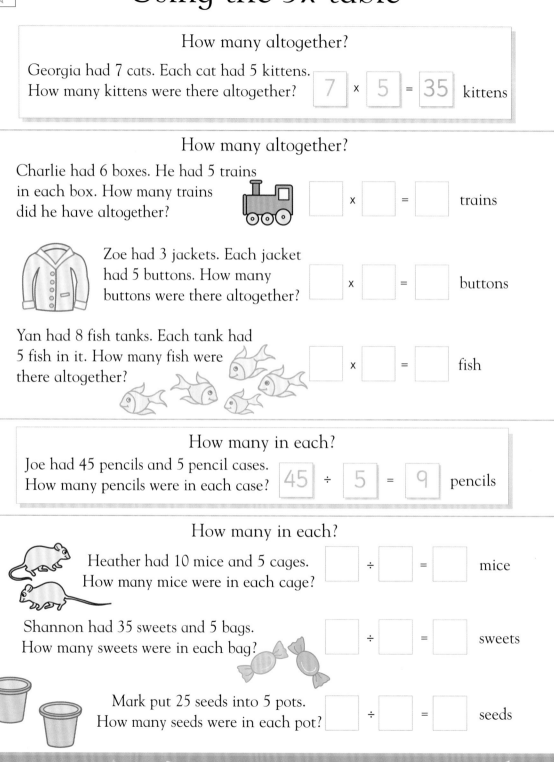

How many altogether?

Georgia had 7 cats. Each cat had 5 kittens.
How many kittens were there altogether?

$\boxed{7}$ x $\boxed{5}$ = $\boxed{35}$ kittens

How many altogether?

Charlie had 6 boxes. He had 5 trains
in each box. How many trains
did he have altogether?

$\boxed{}$ x $\boxed{}$ = $\boxed{}$ trains

Zoe had 3 jackets. Each jacket
had 5 buttons. How many
buttons were there altogether?

$\boxed{}$ x $\boxed{}$ = $\boxed{}$ buttons

Yan had 8 fish tanks. Each tank had
5 fish in it. How many fish were
there altogether?

$\boxed{}$ x $\boxed{}$ = $\boxed{}$ fish

How many in each?

Joe had 45 pencils and 5 pencil cases.
How many pencils were in each case?

$\boxed{45}$ ÷ $\boxed{5}$ = $\boxed{9}$ pencils

How many in each?

Heather had 10 mice and 5 cages.
How many mice were in each cage?

$\boxed{}$ ÷ $\boxed{}$ = $\boxed{}$ mice

Shannon had 35 sweets and 5 bags.
How many sweets were in each bag?

$\boxed{}$ ÷ $\boxed{}$ = $\boxed{}$ sweets

Mark put 25 seeds into 5 pots.
How many seeds were in each pot?

$\boxed{}$ ÷ $\boxed{}$ = $\boxed{}$ seeds

10x table

Count in 10s, colour, and find a pattern.

1	2	3	4	5	6	7	8	9	10
11	12	13	14	15	16	17	18	19	20
21	22	23	24	25	26	27	28	29	30
31	32	33	34	35	36	37	38	39	40
41	42	43	44	45	46	47	48	49	50
51	52	53	54	55	56	57	58	59	60
61	62	63	64	65	66	67	68	69	70
71	72	73	74	75	76	77	78	79	80
81	82	83	84	85	86	87	88	89	90
91	92	93	94	95	96	97	98	99	100

Write the answers.

1 x 10 = 10   2 x 10 = ☐   3 x 10 = ☐   4 x 10 = ☐

5 x 10 = ☐   6 x 10 = ☐   7 x 10 = ☐   8 x 10 = ☐

9 x 10 = ☐   10 x 10 = ☐   11 x 10 = ☐   12 x 10 = ☐

Each box contains 10 crayons. How many crayons are there altogether?

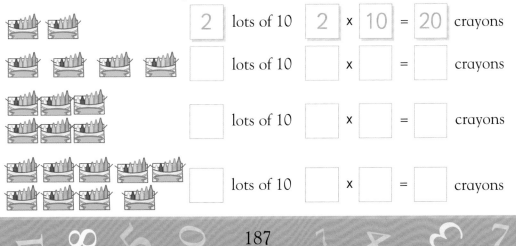

2 lots of 10   2 x 10 = 20 crayons

☐ lots of 10   ☐ x ☐ = ☐ crayons

☐ lots of 10   ☐ x ☐ = ☐ crayons

☐ lots of 10   ☐ x ☐ = ☐ crayons

Multiplying and dividing

Each pod contains 10 peas. How many peas are there altogether?

How many pods? 2

2 × 10 = 20 peas

Work out how many peas.

How many pods? ☐

☐ × 10 = ☐ peas

How many pods? ☐

☐ × ☐ = ☐ peas

How many pods? ☐

☐ × ☐ = ☐ peas

How many pods? ☐

☐ × ☐ = ☐ peas

How many pods did the peas come from?

30

30 ÷ 10 = 3 pods

Work out how many pods.

10

☐ ÷ 10 = ☐ pod

100

☐ ÷ 10 = ☐ pods

20

☐ ÷ 10 = ☐ pods

70

☐ ÷ 10 = ☐ pods

Dividing by 10

One pound is the same as ten 10p coins.

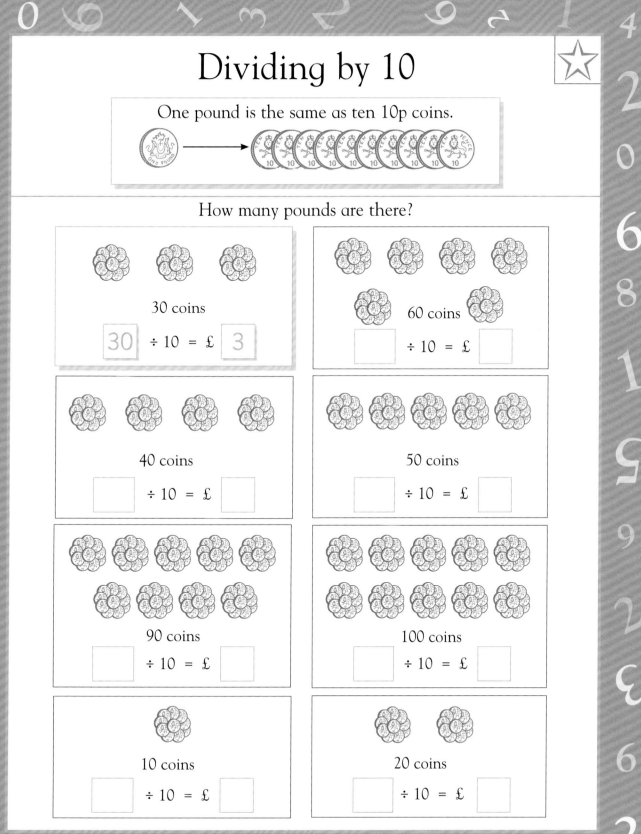

How many pounds are there?

30 coins

30 ÷ 10 = £ 3

60 coins

☐ ÷ 10 = £ ☐

40 coins

☐ ÷ 10 = £ ☐

50 coins

☐ ÷ 10 = £ ☐

90 coins

☐ ÷ 10 = £ ☐

100 coins

☐ ÷ 10 = £ ☐

10 coins

☐ ÷ 10 = £ ☐

20 coins

☐ ÷ 10 = £ ☐

Using the 10x table

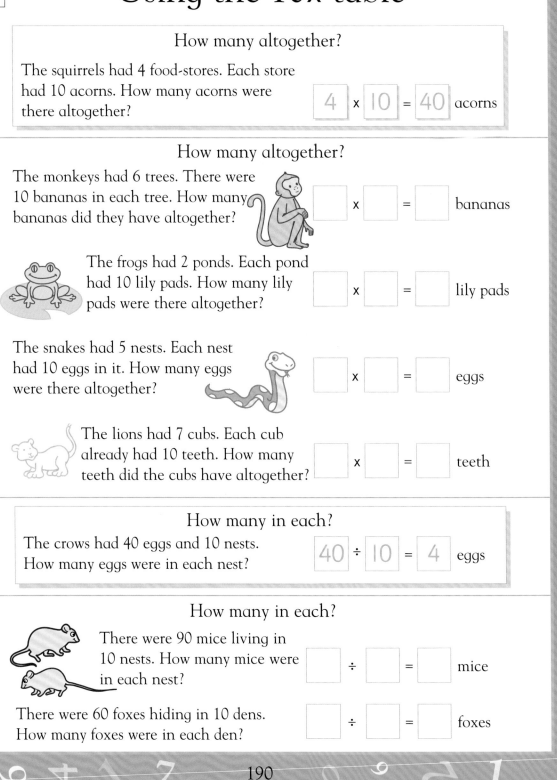

How many altogether?

The squirrels had 4 food-stores. Each store had 10 acorns. How many acorns were there altogether?

$$4 \times 10 = 40 \text{ acorns}$$

How many altogether?

The monkeys had 6 trees. There were 10 bananas in each tree. How many bananas did they have altogether?

☐ x ☐ = ☐ bananas

The frogs had 2 ponds. Each pond had 10 lily pads. How many lily pads were there altogether?

☐ x ☐ = ☐ lily pads

The snakes had 5 nests. Each nest had 10 eggs in it. How many eggs were there altogether?

☐ x ☐ = ☐ eggs

The lions had 7 cubs. Each cub already had 10 teeth. How many teeth did the cubs have altogether?

☐ x ☐ = ☐ teeth

How many in each?

The crows had 40 eggs and 10 nests. How many eggs were in each nest?

$$40 \div 10 = 4 \text{ eggs}$$

How many in each?

There were 90 mice living in 10 nests. How many mice were in each nest?

☐ ÷ ☐ = ☐ mice

There were 60 foxes hiding in 10 dens. How many foxes were in each den?

☐ ÷ ☐ = ☐ foxes

Using the 10x table

Match each dog to the right bone.

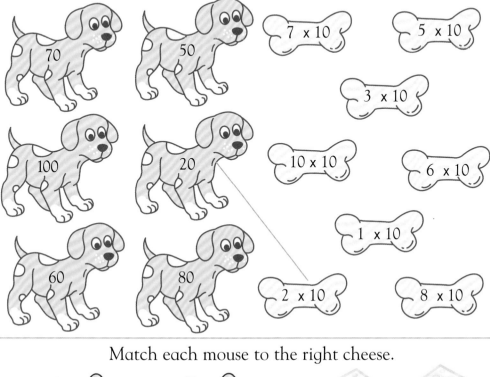

Dogs: 70, 50, 100, 20, 60, 80

Bones: 7 x 10, 5 x 10, 3 x 10, 10 x 10, 6 x 10, 1 x 10, 2 x 10, 8 x 10

Match each mouse to the right cheese.

Mice: 100 ÷ 10, 20 ÷ 10, 70 ÷ 10, 80 ÷ 10, 10 ÷ 10, 50 ÷ 10

Cheese: 7, 12, 2, 10, 5, 8, 1, 3

Using the 10x table

Write in the missing numbers.

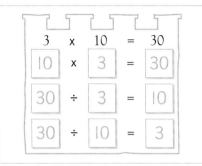

$$3 \times 10 = 30$$
$$10 \times 3 = 30$$
$$30 \div 3 = 10$$
$$30 \div 10 = 3$$

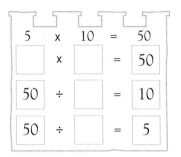

$$5 \times 10 = 50$$
$$\square \times \square = 50$$
$$50 \div \square = 10$$
$$50 \div \square = 5$$

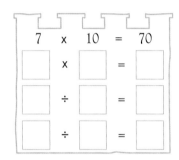

$$7 \times 10 = 70$$
$$\square \times \square = \square$$
$$\square \div \square = \square$$
$$\square \div \square = \square$$

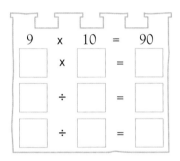

$$9 \times 10 = 90$$
$$\square \times \square = \square$$
$$\square \div \square = \square$$
$$\square \div \square = \square$$

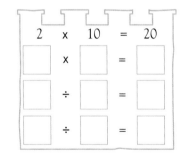

$$2 \times 10 = 20$$
$$\square \times \square = \square$$
$$\square \div \square = \square$$
$$\square \div \square = \square$$

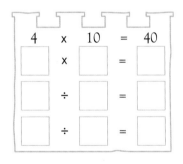

$$4 \times 10 = 40$$
$$\square \times \square = \square$$
$$\square \div \square = \square$$
$$\square \div \square = \square$$

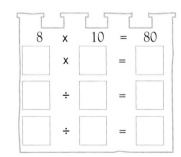

$$8 \times 10 = 80$$
$$\square \times \square = \square$$
$$\square \div \square = \square$$
$$\square \div \square = \square$$

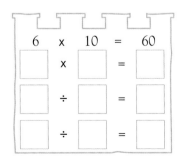

$$6 \times 10 = 60$$
$$\square \times \square = \square$$
$$\square \div \square = \square$$
$$\square \div \square = \square$$

3x table

Count in 3s, colour, and find a pattern.

1	2	3	4	5
6	7	8	9	10
11	12	13	14	15
16	17	18	19	20
21	22	23	24	25

Write the answers.

1 x 3 = 3   2 x 3 =   3 x 3 =   4 x 3 =   5 x 3 =

How many flowers?

2 lots of 3   2 x 3 = 6

lots of 3   x   =

lots of 3   x   =

lots of 3   x   =

193

Multiplying by 3

Write the number sentences to match the pictures.

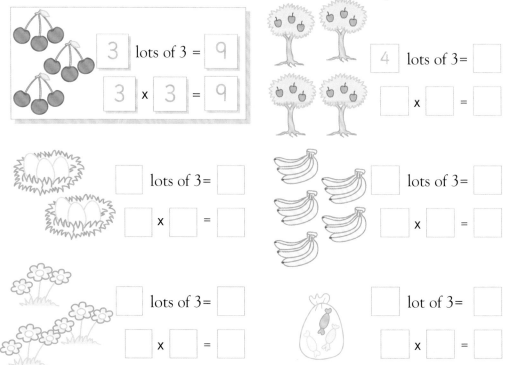

3 lots of 3 = 9

3 × 3 = 9

4 lots of 3 = ☐

☐ × ☐ = ☐

☐ lots of 3 = ☐

☐ × ☐ = ☐

☐ lots of 3 = ☐

☐ × ☐ = ☐

☐ lots of 3 = ☐

☐ × ☐ = ☐

☐ lot of 3 = ☐

☐ × ☐ = ☐

Draw your own pictures to match these number sentences.

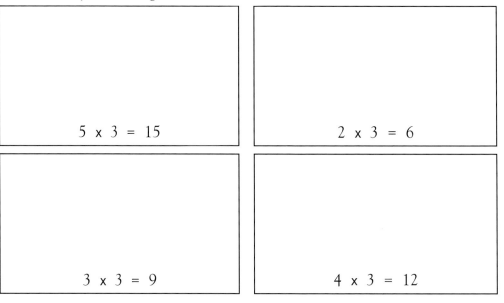

5 × 3 = 15

2 × 3 = 6

3 × 3 = 9

4 × 3 = 12

Dividing by 3

Share the money equally between the purses.
Write a sum to show what you have done.
You might find it easier to change all the money into 1p coins.

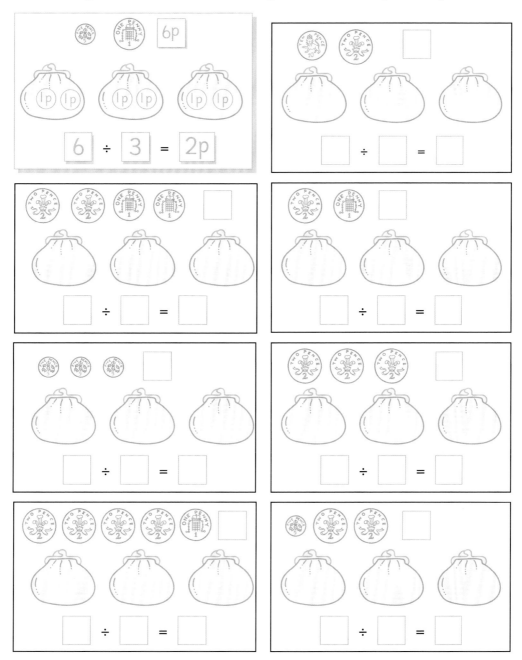

4x table

Count in 4s, colour, and find a pattern.

1	2	3	4	5
6	7	8	9	10
11	12	13	14	15
16	17	18	19	20
21	22	23	24	25

Write the answers.

1 x 4 = ☐4☐   2 x 4 = ☐   3 x 4 = ☐   4 x 4 = ☐   5 x 4 = ☐

How many flowers?

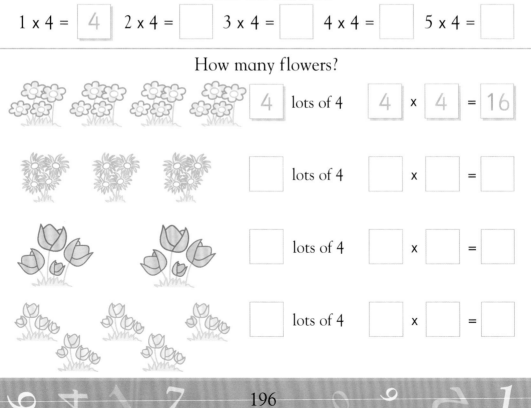

☐4☐ lots of 4   ☐4☐ x ☐4☐ = ☐16☐

☐ lots of 4   ☐ x ☐ = ☐

☐ lots of 4   ☐ x ☐ = ☐

☐ lots of 4   ☐ x ☐ = ☐

Multiplying by 4

Write number sentences to match the pictures.

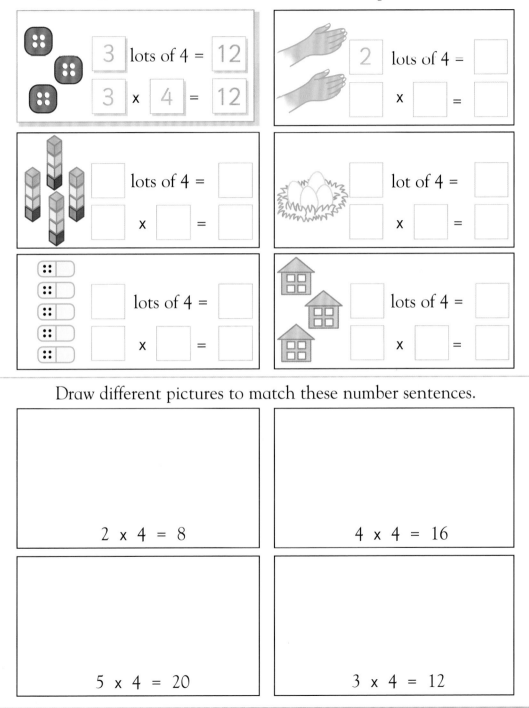

3 lots of 4 = 12

3 x 4 = 12

2 lots of 4 = ☐

☐ x ☐ = ☐

☐ lots of 4 = ☐

☐ x ☐ = ☐

☐ lot of 4 = ☐

☐ x ☐ = ☐

☐ lots of 4 = ☐

☐ x ☐ = ☐

☐ lots of 4 = ☐

☐ x ☐ = ☐

Draw different pictures to match these number sentences.

2 x 4 = 8

4 x 4 = 16

5 x 4 = 20

3 x 4 = 12

Dividing by 4

How many on each plate?

There are 4 children. How many things will each child have?
Draw the objects in the circles.

8 sandwiches

8 ÷ 4 = 2 each

12 biscuits

☐ ÷ 4 = ☐ each

4 drinks

☐ ÷ ☐ = ☐ each

20 cherries

☐ ÷ ☐ = ☐ each

16 cakes

☐ ÷ ☐ = ☐ each

8 cheese triangles

☐ ÷ ☐ = ☐ each

Mixed tables

How many pegs are there in each pegboard? Write the sum.

3 rows of 4

3 x 4 = 12

How many pegs are there in each pegboard? Write the sums.

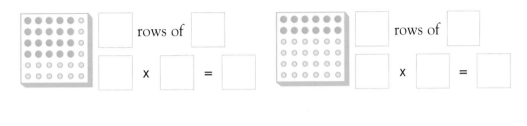

_____ rows of _____

_____ x _____ = _____

_____ rows of _____

_____ x _____ = _____

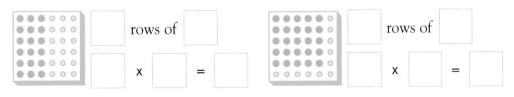

_____ rows of _____

_____ x _____ = _____

_____ rows of _____

_____ x _____ = _____

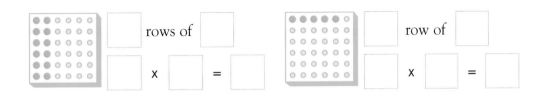

_____ rows of _____

_____ x _____ = _____

_____ row of _____

_____ x _____ = _____

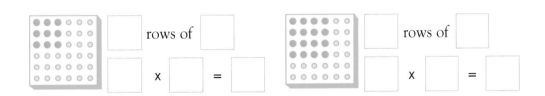

_____ rows of _____

_____ x _____ = _____

_____ rows of _____

_____ x _____ = _____

Mixed tables

Share the 12 pennies equally. Draw the coins
and write the sum to show how many each person gets.

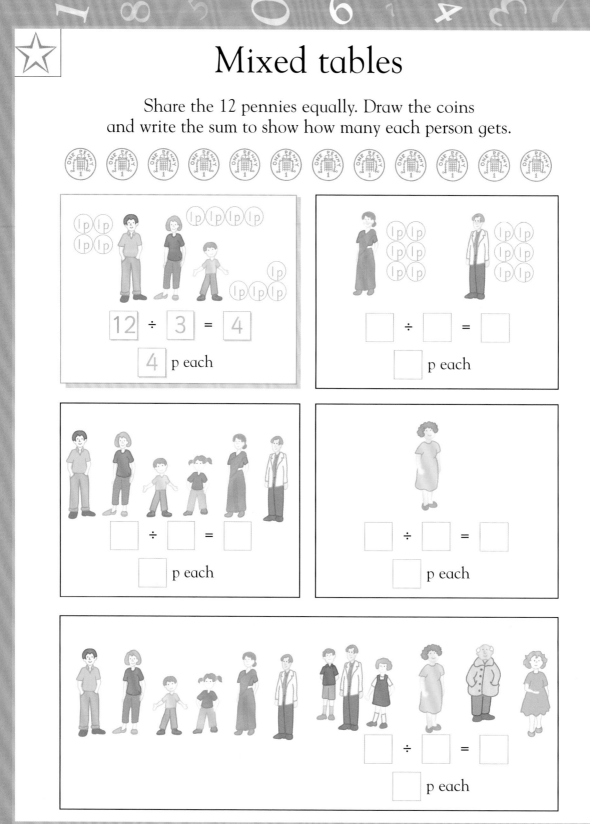

12 ÷ 3 = 4

4 p each

☐ ÷ ☐ = ☐

☐ p each

☐ ÷ ☐ = ☐

☐ p each

☐ ÷ ☐ = ☐

☐ p each

☐ ÷ ☐ = ☐

☐ p each

Mixed tables

How much will they get paid?

Price List for Jobs
Dust bedroom	3p
Feed rabbit	2p
Tidy toys	6p
Fetch newspaper	5p
Walk dog	10p

Write a sum to show how much money Joe and Jasmine will get for these jobs.

Feed 4 rabbits   4 x 2p = 8p

Dust 2 bedrooms   ☐ x ☐ = ☐ p

Walk the dog 4 times   ☐ x ☐ = ☐ p

Tidy the toys 3 times   ☐ x ☐ = ☐ p

Fetch the newspaper 5 times   ☐ x ☐ = ☐ p

How much will they get for these jobs?
Use the space for your working out.

Dust 3 bedrooms and walk the dog twice   ☐ + ☐ = ☐ p

Feed the rabbit 10 times and tidy the toys twice   ☐ + ☐ = ☐ p

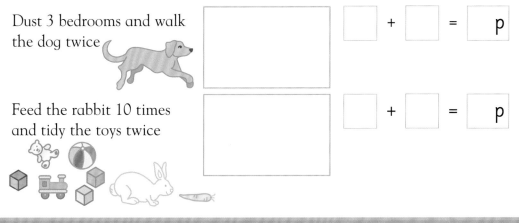

Mixed tables

Write the numbers that the raindrops are hiding.

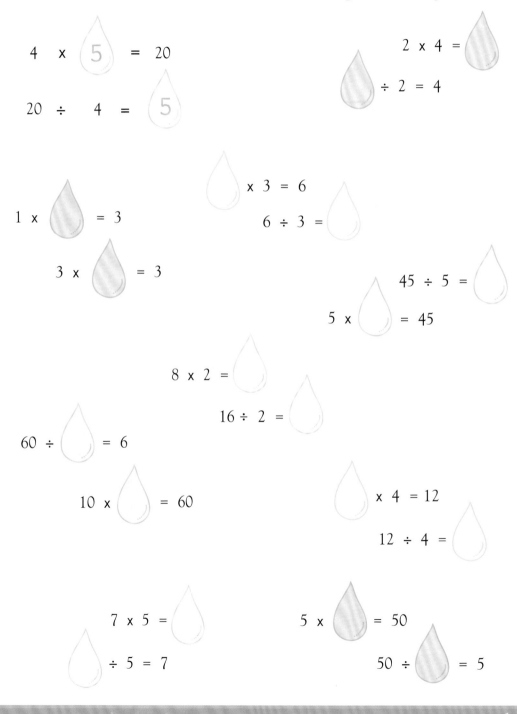

4 x 5 = 20

20 ÷ 4 = 5

2 x 4 =

÷ 2 = 4

1 x = 3

x 3 = 6

6 ÷ 3 =

3 x = 3

45 ÷ 5 =

5 x = 45

8 x 2 =

16 ÷ 2 =

60 ÷ = 6

10 x = 60

x 4 = 12

12 ÷ 4 =

7 x 5 =

÷ 5 = 7

5 x = 50

50 ÷ = 5

Mixed tables

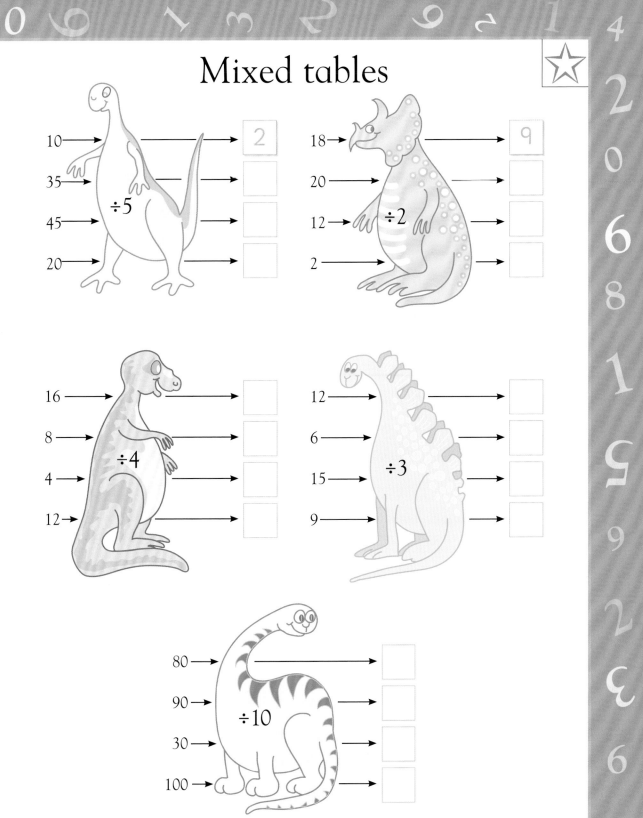

10 → ÷5 → 2

35 → →

45 → →

20 → →

18 → ÷2 → 9

20 → →

12 → →

2 → →

16 → ÷4 →

8 → →

4 → →

12 → →

12 → ÷3 →

6 → →

15 → →

9 → →

80 → ÷10 →

90 → →

30 → →

100 → →

Mixed tables

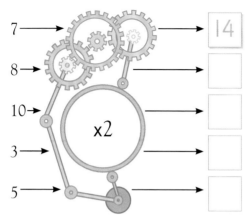

7 → ⟶ 14
8 → ⟶ ☐
10 → x2 ⟶ ☐
3 → ⟶ ☐
5 → ⟶ ☐

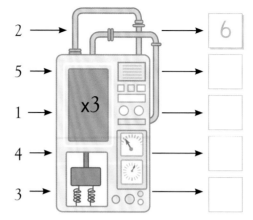

2 → ⟶ 6
5 → ⟶ ☐
1 → x3 ⟶ ☐
4 → ⟶ ☐
3 → ⟶ ☐

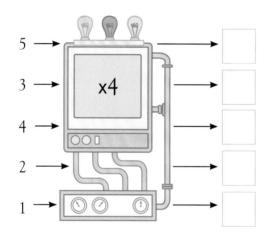

5 → ⟶ ☐
3 → x4 ⟶ ☐
4 → ⟶ ☐
2 → ⟶ ☐
1 → ⟶ ☐

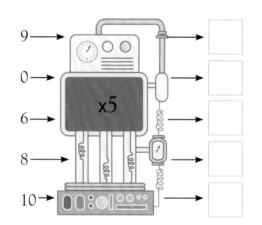

9 → ⟶ ☐
0 → ⟶ ☐
6 → x5 ⟶ ☐
8 → ⟶ ☐
10 → ⟶ ☐

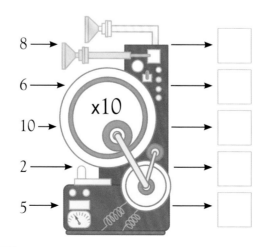

8 → ⟶ ☐
6 → ⟶ ☐
10 → x10 ⟶ ☐
2 → ⟶ ☐
5 → ⟶ ☐

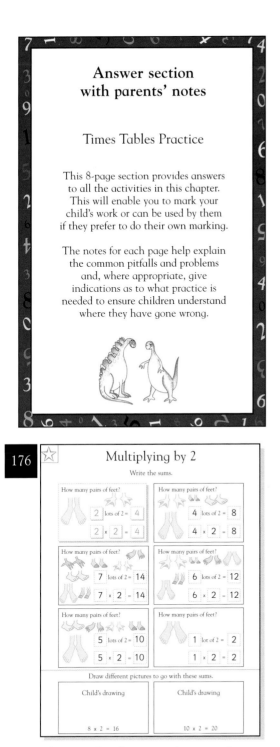

175

2x table

Count in 2s, colour, and find a pattern.

1	2	3	4	5
6	7	8	9	10
11	12	13	14	15
16	17	18	19	20
21	22	23	24	25

Write the answers.

$1 \times 2 = 2$ $2 \times 2 = 4$ $3 \times 2 = 6$ $4 \times 2 = 8$

$5 \times 2 = 10$ $6 \times 2 = 12$ $7 \times 2 = 14$ $8 \times 2 = 16$

$9 \times 2 = 18$ $10 \times 2 = 20$ $11 \times 2 = 22$ $12 \times 2 = 24$

How many ears?

5 lots of 2 $5 \times 2 = 10$ ears

2 lots of 2 $2 \times 2 = 4$ ears

8 lots of 2 $8 \times 2 = 16$ ears

4 lots of 2 $4 \times 2 = 8$ ears

Children should realise that multiplying by 2 is really the same as adding 2 repeatedly. 2×3 is the same as $2+2+2$. Help them realise that the pattern they have coloured has all even numbers. Can they use this to tell you whether 27 or 31 is in the $2 \times$ table?

176

Multiplying by 2

Write the sums.

How many pairs of feet?
2 lots of 2 = 4
$2 \times 2 = 4$

How many pairs of feet?
4 lots of 2 = 8
$4 \times 2 = 8$

How many pairs of feet?
7 lots of 2 = 14
$7 \times 2 = 14$

How many pairs of feet?
6 lots of 2 = 12
$6 \times 2 = 12$

How many pairs of feet?
5 lots of 2 = 10
$5 \times 2 = 10$

How many pairs of feet?
1 lot of 2 = 2
$1 \times 2 = 2$

Draw different pictures to go with these sums.

Child's drawing	Child's drawing
$8 \times 2 = 16$	$10 \times 2 = 20$

Can children say what they see in the picture (e.g. 3 lots of 2 feet) before they read the number sentence? Ask children to set out objects like building bricks in 'lots' of 2.

177

Dividing by 2

Share the eggs equally between the nests.

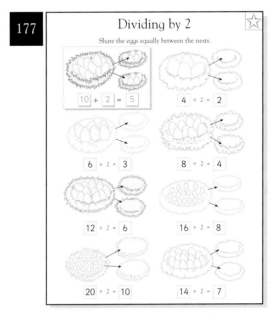

$10 \div 2 = 5$ $4 \div 2 = 2$

$6 \div 2 = 3$ $8 \div 2 = 4$

$12 \div 2 = 6$ $16 \div 2 = 8$

$20 \div 2 = 10$ $14 \div 2 = 7$

Do children know that the ÷ sign means sharing things out into equal groups or piles? Small buttons can represent the eggs and they can actually share these out between the two smaller nests each time. At this stage lots of practical work is important.

Using the 2x table

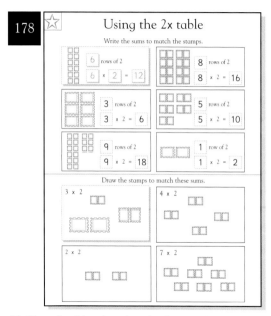

Write the sums to match the stamps.

6 rows of 2
6 x 2 = 12

8 rows of 2
8 x 2 = 16

3 rows of 2
3 x 2 = 6

5 rows of 2
5 x 2 = 10

9 rows of 2
9 x 2 = 18

1 row of 2
1 x 2 = 2

Draw the stamps to match these sums.

3 x 2   4 x 2

2 x 2   7 x 2

Children should realise that the first number (the number of rows) is counted down the strip and not across. Explain how using their times tables and counting the rows and the number across saves them time. This task will be much easier if children can already recite their 2x table.

Using the 2x table

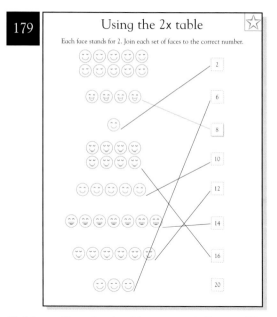

Each face stands for 2. Join each set of faces to the correct number.

2
6
8
10
12
14
16
20

Children will need to be able to count in 2s up to 20 before tackling this activity. If they find it difficult to see each single face as a '2', it would be worth using a collection of 2p coins and reinforcing the fact that there is only one coin but it represents two pence.

Using the 2x table

How many eyes?

3 x 2 = 6 eyes   5 x 2 = 10 eyes

9 x 2 = 18 eyes   2 x 2 = 4 eyes

8 x 2 = 16 eyes   4 x 2 = 8 eyes

Draw your own pictures to match these number sentences.

2 x 2 = 4   10 x 2 = 20
              Child's drawing

3 x 2 = 6   7 x 2 = 14
Child's drawing   Child's drawing

Encourage your child to talk out loud as they are doing the activity. Then you will be able to check that they understand what they are doing. Are they using the number of pairs as the first number and the number in the pair (i.e. 2) as the second?

5x table

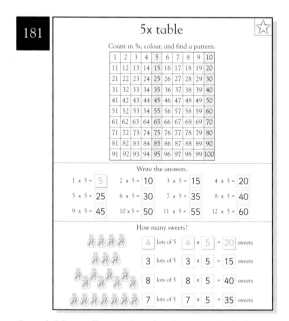

Count in 5s, colour, and find a pattern.

1	2	3	4	5	6	7	8	9	10
11	12	13	14	15	16	17	18	19	20
21	22	23	24	25	26	27	28	29	30
31	32	33	34	35	36	37	38	39	40
41	42	43	44	45	46	47	48	49	50
51	52	53	54	55	56	57	58	59	60
61	62	63	64	65	66	67	68	69	70
71	72	73	74	75	76	77	78	79	80
81	82	83	84	85	86	87	88	89	90
91	92	93	94	95	96	97	98	99	100

Write the answers.

1 x 5 = 5   2 x 5 = 10   3 x 5 = 15   4 x 5 = 20

5 x 5 = 25   6 x 5 = 30   7 x 5 = 35   8 x 5 = 40

9 x 5 = 45   10 x 5 = 50   11 x 5 = 55   12 x 5 = 60

How many sweets?

4 lots of 5   4 x 5 = 20 sweets

3 lots of 5   3 x 5 = 15 sweets

8 lots of 5   8 x 5 = 40 sweets

7 lots of 5   7 x 5 = 35 sweets

Can children tell you what they have noticed about the numbers they have coloured in on the grid? (The final digit is always 0 or 5.) Can they use this information to tell you whether 78, 90, or 23, or even 995 are in the 5x table?

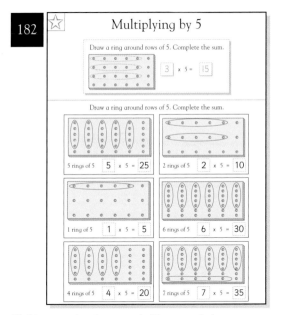

182 Multiplying by 5

Draw a ring around rows of 5. Complete the sum.

3 x 5 = 15

Draw a ring around rows of 5. Complete the sum.

5 rings of 5 5 x 5 = 25

2 rings of 5 2 x 5 = 10

1 ring of 5 1 x 5 = 5

6 rings of 5 6 x 5 = 30

4 rings of 5 4 x 5 = 20

7 rings of 5 7 x 5 = 35

Children need to be reminded how much faster it is to be able to say, '7 rows of 5, that's 7x5, that's 35' than it would be to count all 35 holes individually.

183 Dividing by 5

Write a number sentence to show how many cubes are in each tower.

15 cubes altogether

5 towers

15 ÷ 5 = 3

Write a number sentence to show how many cubes are in each tower.

20 cubes altogether
5 towers
20 ÷ 5 = 4

30 cubes altogether
5 towers
30 ÷ 5 = 6

25 cubes altogether
5 towers
25 ÷ 5 = 5

10 cubes altogether
5 towers
10 ÷ 5 = 2

35 cubes altogether
5 towers
35 ÷ 5 = 7

40 cubes altogether
5 towers
40 ÷ 5 = 8

Using building bricks to show each sum practically would reinforce understanding. Can children describe in words what they have done, e.g. '40 bricks shared between 5 towers gives 8 bricks in each tower'?

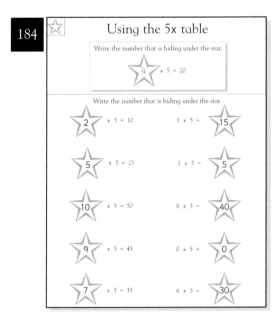

184 Using the 5x table

Write the number that is hiding under the star.

4 x 5 = 20

Write the number that is hiding under the star.

2 x 5 = 10

3 x 5 = 15

5 x 5 = 25

1 x 5 = 5

10 x 5 = 50

8 x 5 = 40

9 x 5 = 45

0 x 5 = 0

7 x 5 = 35

6 x 5 = 30

Children need to be able to recite the 5x table (even if only slowly and in order) before they tackle this activity. Encourage them to read the number sentence out loud: 'something (the star) times 5 = 10', before they try to work out what the 'something' is.

185 Using the 5x table

Each frog stands for 5. Join each set of frogs to the correct number.

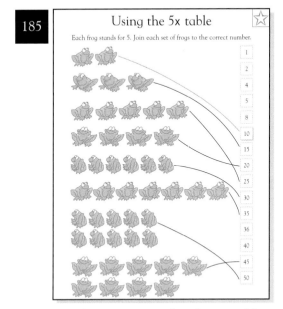

| 1 |
| 2 |
| 4 |
| 5 |
| 8 |
| 10 |
| 15 |
| 20 |
| 25 |
| 30 |
| 35 |
| 36 |
| 40 |
| 45 |
| 50 |

Explain that there are more numbers than sets of frogs, and therefore not every number will join up. Remind children that the last digit in all the numbers in the 5x table is either 0 or 5. Can they use this fact to predict which numbers will not be joined to the frogs?

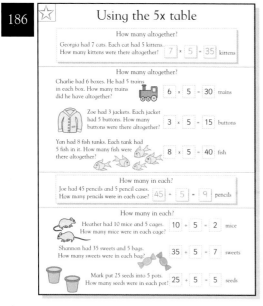

186 ☆ Using the 5x table

How many altogether?

Georgia had 7 cats. Each cat had 5 kittens.
How many kittens were there altogether? $7 \times 5 = 35$ kittens

How many altogether?

Charlie had 6 boxes. He had 5 trains in each box. How many trains did he have altogether? $6 \times 5 = 30$ trains

Zoe had 3 jackets. Each jacket had 5 buttons. How many buttons were there altogether? $3 \times 5 = 15$ buttons

Yan had 8 fish tanks. Each tank had 5 fish in it. How many fish were there altogether? $8 \times 5 = 40$ fish

How many in each?

Joe had 45 pencils and 5 pencil cases. How many pencils were in each case? $45 \div 5 = 9$ pencils

How many in each?

Heather had 10 mice and 5 cages. How many mice were in each cage? $10 \div 5 = 2$ mice

Shannon had 35 sweets and 5 bags. How many sweets were in each bag? $35 \div 5 = 7$ sweets

Mark put 25 seeds into 5 pots. How many seeds were in each pot? $25 \div 5 = 5$ seeds

Children might like to draw each situation to help them visualise it, or they might like to use counters (pasta shapes) to represent the objects.

187 ☆ 10x table

Count in 10s, colour, and find a pattern.

1	2	3	4	5	6	7	8	9	10
11	12	13	14	15	16	17	18	19	20
21	22	23	24	25	26	27	28	29	30
31	32	33	34	35	36	37	38	39	40
41	42	43	44	45	46	47	48	49	50
51	52	53	54	55	56	57	58	59	60
61	62	63	64	65	66	67	68	69	70
71	72	73	74	75	76	77	78	79	80
81	82	83	84	85	86	87	88	89	90
91	92	93	94	95	96	97	98	99	100

Write the answers.

$1 \times 10 = 10$ $2 \times 10 = 20$ $3 \times 10 = 30$ $4 \times 10 = 40$

$5 \times 10 = 50$ $6 \times 10 = 60$ $7 \times 10 = 70$ $8 \times 10 = 80$

$9 \times 10 = 90$ $10 \times 10 = 100$ $11 \times 10 = 110$ $12 \times 10 = 120$

Each box contains 10 crayons. How many crayons are there altogether?

2 lots of 10 $2 \times 10 = 20$ crayons

4 lots of 10 $4 \times 10 = 40$ crayons

6 lots of 10 $6 \times 10 = 60$ crayons

9 lots of 10 $9 \times 10 = 90$ crayons

What do children notice about the numbers in the sequence? (The tens go up in ones while the units are always 0.) Ask them to use what they know to predict whether other numbers, such as 74, 12543, and 990, are in the sequence of tens.

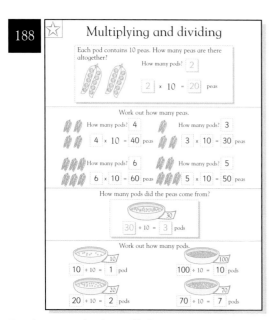

188 ☆ Multiplying and dividing

Each pod contains 10 peas. How many peas are there altogether?

How many pods? 2

$2 \times 10 = 20$ peas

Work out how many peas.

How many pods? 4 $4 \times 10 = 40$ peas

How many pods? 3 $3 \times 10 = 30$ peas

How many pods? 6 $6 \times 10 = 60$ peas

How many pods? 5 $5 \times 10 = 50$ peas

How many pods did the peas come from?

$30 \div 10 = 3$ pods

Work out how many pods.

$10 \div 10 = 1$ pod $100 \div 10 = 10$ pods

$20 \div 10 = 2$ pods $70 \div 10 = 7$ pods

Dividing a number by itself often causes confusion. It would be worth getting 10 dried peas (or some other counter) and letting children share them out between 10 cups (or rings drawn on paper) so that they can actually see that 10÷10 will only give 1 in each set.

189 ☆ Dividing by 10

One pound is the same as ten 10p coins.

How many pounds are there?

30 coins $30 \div 10 = £3$

60 coins $60 \div 10 = £6$

40 coins $40 \div 10 = £4$

50 coins $50 \div 10 = £5$

90 coins $90 \div 10 = £9$

100 coins $100 \div 10 = £10$

10 coins $10 \div 10 = £1$

20 coins $20 \div 10 = £2$

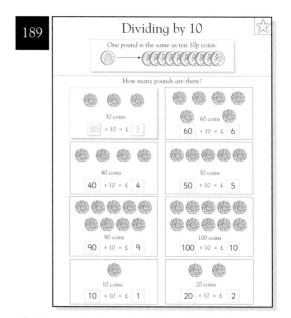

Children need to know their 10x table for this page, as it is not practical to simply count the coins out. They should realise that if they know that 8x (lots of) 10 is 80, then they also know that 80÷ (shared by) 10 is 8. These related facts ought to be reinforced.

Using the 10x table

How many altogether?
The squirrels had 4 food-stores. Each store had 10 acorns. How many acorns were there altogether? $4 \times 10 = 40$ acorns

How many altogether?
The monkeys had 6 trees. There were 10 bananas in each tree. How many bananas did they have altogether? $6 \times 10 = 60$ bananas

The frogs had 2 ponds. Each pond had 10 lily pads. How many lily pads were there altogether? $2 \times 10 = 20$ lily pads

The snakes had 5 nests. Each nest had 10 eggs in it. How many eggs were there altogether? $5 \times 10 = 50$ eggs

The lions had 7 cubs. Each cub already had 10 teeth. How many teeth did the cubs have altogether? $7 \times 10 = 70$ teeth

How many in each?
The crows had 40 eggs and 10 nests. How many eggs were in each nest? $40 \div 10 = 4$ eggs

How many in each?
There were 90 mice living in 10 nests. How many mice were in each nest? $90 \div 10 = 9$ mice

There were 60 foxes hiding in 10 dens. How many foxes were in each den? $60 \div 10 = 6$ foxes

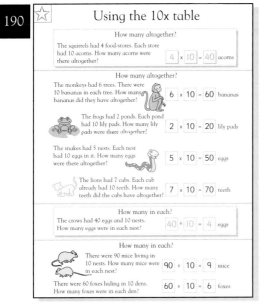

Can children see that these numbers are too large for them to draw or to use objects to 'act out' the problems? They will need to use their times tables facts to calculate the answers in their heads.

Using the 10x table

Match each dog to the right bone.

Match each mouse to the right cheese.

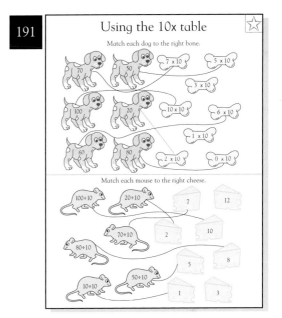

Ask children to explain what 'x' and '÷' mean. Not all pictures join up. Some children see a need to join everything to something, and have been caught out in end of Key Stage tests for doing this. This page gives practice in only joining when there is a match.

Using the 10x table

Write in the missing numbers.

3	x	10	=	30
10	x	3	=	30
30	÷	3	=	10
30	÷	10	=	3

5	x	10	=	50
10	x	5	=	50
50	÷	5	=	10
50	÷	10	=	5

7	x	10	=	70
10	x	7	=	70
70	÷	7	=	10
70	÷	10	=	7

9	x	10	=	90
10	x	9	=	90
90	÷	9	=	10
90	÷	10	=	9

2	x	10	=	20
10	x	2	=	20
20	÷	2	=	10
20	÷	10	=	2

4	x	10	=	40
10	x	4	=	40
40	÷	4	=	10
40	÷	10	=	4

8	x	10	=	80
10	x	8	=	80
80	÷	8	=	10
80	÷	10	=	8

6	x	10	=	60
10	x	6	=	60
60	÷	6	=	10
60	÷	10	=	6

Using practical objects (such as 8 pencils) will help children's understanding of how the whole set can be shared out in different ways.

3x table

Count in 3s, colour, and find a pattern.

1	2	3	4	5
6	7	8	9	10
11	12	13	14	15
16	17	18	19	20
21	22	23	24	25

Write the answers.
$1 \times 3 = 3$ $2 \times 3 = 6$ $3 \times 3 = 9$ $4 \times 3 = 12$ $5 \times 3 = 15$

How many flowers?

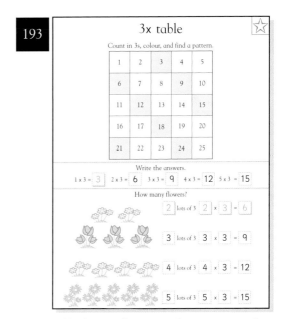

2 lots of 3 $2 \times 3 = 6$

3 lots of 3 $3 \times 3 = 9$

4 lots of 3 $4 \times 3 = 12$

5 lots of 3 $5 \times 3 = 15$

The 3x table does not have very obvious patterns for young children to identify, which is why it tends to be taught only up to 5x 3 at average infant level. If children are able to use and learn the table beyond this level, then do not hold them back.

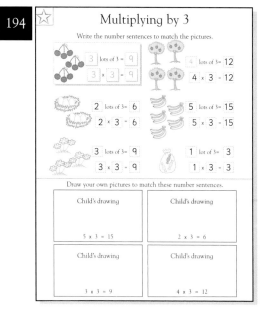

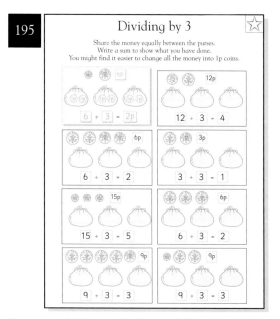

Encourage children to use mathematical language in sentences such as '5 lots of 3 bananas is 15 bananas altogether'. This will help them reinforce their understanding of what the written symbols mean and will also help you to check on that understanding.

Encourage children to double-check the amount of money in each purse. Using 1p coins will be helpful. Do they realise that the amounts in some questions are the same, although the coins are different? Thus the amount shared out to each purse is the same.

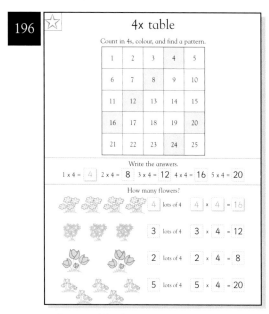

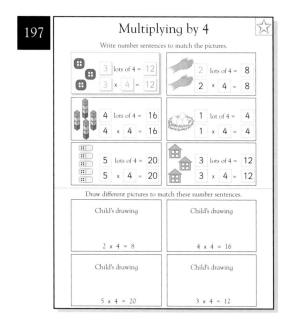

Learning the 4x table above the five-times level is not essential at this stage. However, if children are sure of the 2s, 5s, and 10s, then help them to learn beyond. The number pattern is not easy to predict, but they may notice that all answers are even numbers.

This page provides further practice in the 4x table.

198 — Dividing by 4

How many on each plate?

There are 4 children. How many things will each child have?
Draw the objects in the circles.

8 sandwiches
$8 ÷ 4 = 2$ each

12 biscuits
$12 ÷ 4 = 3$ each

4 drinks
$4 ÷ 4 = 1$ each

20 cherries
$20 ÷ 4 = 5$ each

16 cakes
$16 ÷ 4 = 4$ each

8 cheese triangles
$8 ÷ 4 = 2$ each

Let children refer back to the number square on p.196.
They can find 12 on the grid and count back to see how
many lots of 4 it took to reach it. They may set a real
table and talk mathematically about it: 'There are 4
places, I have 8 apples, that is 2 apples each.'

199 — Mixed tables

How many pegs are there in each pegboard? Write the sum.

3 rows of 4
$3 × 4 = 12$

How many pegs are there in each pegboard? Write the sums.

4 rows of 5
$4 × 5 = 20$

2 rows of 6
$2 × 6 = 12$

3 rows of 6
$3 × 6 = 18$

5 rows of 6
$5 × 6 = 30$

6 rows of 2
$6 × 2 = 12$

1 row of 5
$1 × 5 = 5$

3 rows of 3
$3 × 3 = 9$

4 rows of 4
$4 × 4 = 16$

Ask children to find two multiplication sums for each
board. With question 1 they could find $4 × 5 = 20$ and
$5 × 4 = 20$. Multiplication is 'commutative', giving
the same answer whichever way around you put the
numbers. Understanding this will help a lot later on.

200 — Mixed tables

Share the 12 pennies equally. Draw the coins
and write the sum to show how many each person gets.

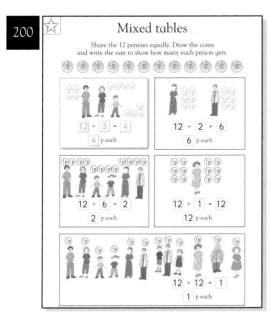

$12 ÷ 3 = 4$
4 p each

$12 ÷ 2 = 6$
6 p each

$12 ÷ 6 = 2$
2 p each

$12 ÷ 1 = 12$
12 p each

$12 ÷ 12 = 1$
1 p each

When working with money it is important that the
correct units are used. Encourage children to say the unit
with each answer, for instance '6p', each time.

201 — Mixed tables

How much will they get paid?

Price List for Jobs
Dust bedroom   3p
Feed rabbit   2p
Tidy toys   6p
Fetch newspaper 5p
Walk dog   10p

Write a sum to show how much money Joe and Jasmine will get for these jobs.

Feed 4 rabbits $4 × 2p = 8p$

Dust 2 bedrooms $2 × 3p = 6p$

Walk the dog 4 times $4 × 10p = 40p$

Tidy the toys 3 times $3 × 6p = 18p$

Fetch the newspaper 5 times $5 × 5p = 25p$

How much will they get for these jobs?
Use the space for your working out.

Dust 3 bedrooms and walk
the dog twice
$3 × 3 = 9$
$2 × 10 = 20$
$9 + 20p = 29p$

Feed the rabbit 10 times
and tidy the toys twice
$10 × 2 = 20$
$2 × 6 = 12$
$20p + 12p = 32p$

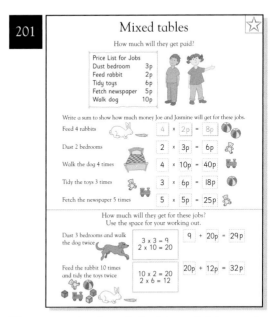

Writing or saying the unit (in this case 'p') is a good
habit for money problems. In the 6th and 7th question,
do children realise that they need to do more than one
sum? Having found the cost of the two jobs separately
they need to remember to add them together.

Mixed tables

Write the numbers that the raindrops are hiding.

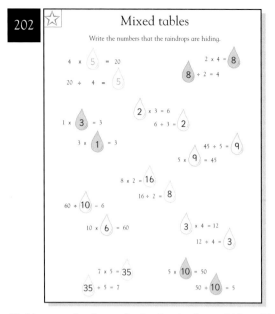

$4 \times 5 = 20$

$20 \div 4 = 5$

$2 \times 4 = 8$

$8 \div 2 = 4$

$1 \times 3 = 3$

$3 \times 1 = 3$

$2 \times 3 = 6$

$6 \div 3 = 2$

$45 \div 5 = 9$

$5 \times 9 = 45$

$8 \times 2 = 16$

$16 \div 2 = 8$

$60 \div 10 = 6$

$10 \times 6 = 60$

$3 \times 4 = 12$

$12 \div 4 = 3$

$7 \times 5 = 35$

$35 \div 5 = 7$

$5 \times 10 = 50$

$50 \div 10 = 5$

Children need to know the 2x, 3x, 4x, 5x, and 10x tables for this page. It is important to know that multiplication and division are 'opposite' (inverse) operations. So they should know that if 4x5=20 and 5x4= 20, then 20÷4=5 and 20÷5=4.

Mixed tables

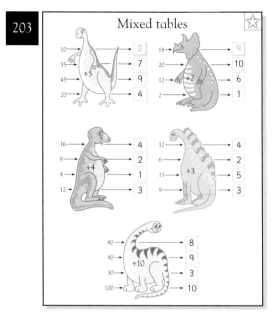

Children should try to do this exercise using purely mental calculations. Can they talk themselves through the exercises using mathematical sentences such as, '35÷5 ... means how many lots of 5 in 35 ...7 fives are 35... so 35÷5=7'?

Mixed tables

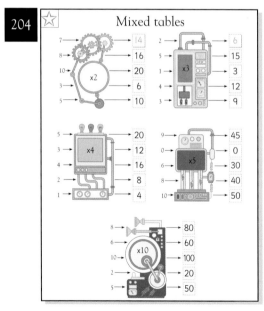

If children know their tables well, they might find it challenging to do this page as a 'race against the clock' by using a kitchen timer, for example. If you cover their answers, they could have a second race and try to beat their own time!

MORE

TiMES

TaBLES

PRACTICE

Author and Consultant Sean McArdle

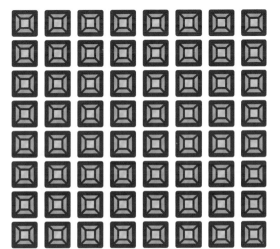

Contents

Speed trials

Write the answers as fast as you can, but get them right!

3 x 2 =	0 x 5 =	3 x 10 =	0 x 3 =
5 x 2 =	10 x 5 =	5 x 10 =	10 x 3 =
1 x 2 =	8 x 5 =	1 x 10 =	8 x 3 =
4 x 2 =	6 x 5 =	4 x 10 =	6 x 3 =
12 x 2 =	2 x 5 =	7 x 10 =	2 x 3 =
2 x 2 =	7 x 5 =	2 x 10 =	7 x 3 =
6 x 2 =	4 x 5 =	6 x 10 =	4 x 3 =
8 x 2 =	1 x 5 =	8 x 10 =	1 x 3 =
10 x 2 =	5 x 5 =	10 x 11 =	5 x 3 =
0 x 2 =	3 x 5 =	0 x 10 =	3 x 3 =
9 x 2 =	5 x 3 =	9 x 10 =	6 x 4 =
2 x 7 =	5 x 8 =	10 x 7 =	3 x 4 =
2 x 11 =	5 x 6 =	10 x 1 =	7 x 12 =
2 x 4 =	5 x 9 =	10 x 4 =	4 x 4 =
3 x 7 =	5 x 7 =	10 x 7 =	10 x 4 =
2 x 5 =	11 x 4 =	10 x 12 =	8 x 4 =
2 x 9 =	5 x 1 =	10 x 9 =	0 x 4 =
2 x 6 =	4 x 7 =	10 x 6 =	9 x 4 =
2 x 8 =	5 x 11 =	10 x 8 =	5 x 4 =
12 x 3 =	5 x 2 =	10 x 3 =	2 x 4 =

All the 3s

You will need to know these:

$1 \times 3 = 3$ $2 \times 3 = 6$ $3 \times 3 = 9$ $4 \times 3 = 12$ $5 \times 3 = 15$ $10 \times 3 = 30$

How many altogether?

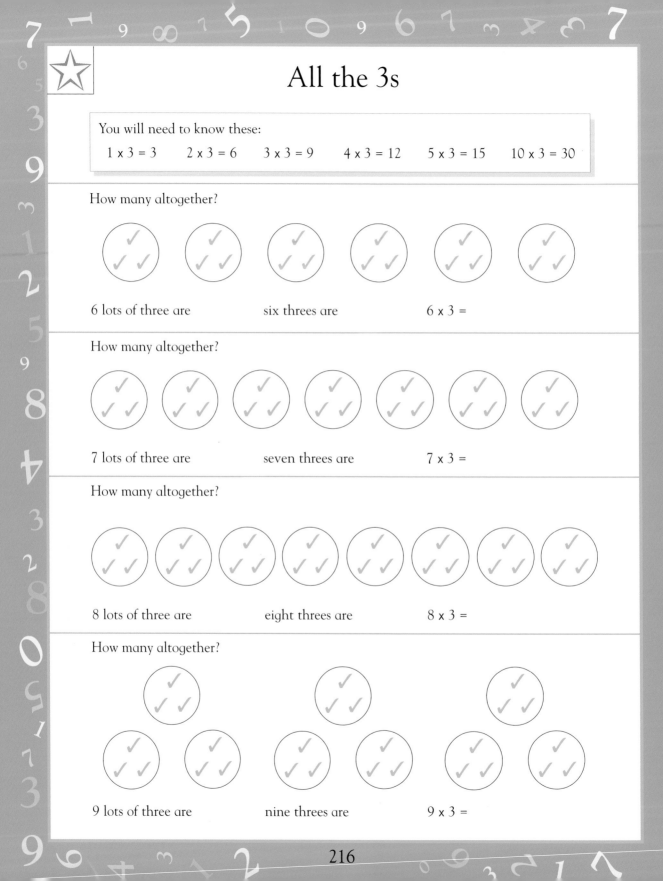

6 lots of three are   six threes are $6 \times 3 =$

How many altogether?

7 lots of three are   seven threes are $7 \times 3 =$

How many altogether?

8 lots of three are   eight threes are $8 \times 3 =$

How many altogether?

9 lots of three are   nine threes are $9 \times 3 =$

All the 3s again

Cover the 3 times table with a piece of paper so you can't see the numbers.
Write the answers. Be as fast as you can, but get them right!

2 x 3 =	5 x 3 =	6 x 3 =
3 x 3 =	7 x 3 =	9 x 3 =
4 x 3 =	9 x 3 =	4 x 3 =
5 x 3 =	4 x 3 =	5 x 3 =
6 x 3 =	6 x 3 =	3 x 7 =
7 x 3 =	8 x 3 =	3 x 4 =
8 x 3 =	10 x 3 =	2 x 3 =
9 x 3 =	11 x 3 =	12 x 3 =
10 x 3 =	12 x 3 =	3 x 9 =
11 x 3 =	2 x 3 =	3 x 6 =
3 x 2 =	3 x 5 =	3 x 5 =
3 x 3 =	3 x 7 =	3 x 8 =
3 x 4 =	3 x 9 =	7 x 3 =
3 x 5 =	3 x 4 =	3 x 2 =
3 x 6 =	3 x 6 =	3 x 11 =
3 x 7 =	3 x 8 =	8 x 3 =
3 x 8 =	3 x 10 =	3 x 10 =
3 x 9 =	3 x 1 =	1 x 3 =
3 x 10 =	3 x 0 =	3 x 3 =
3 x 12 =	3 x 2 =	3 x 9 =

All the 4s

You should know these:

1 x 4 = 4   2 x 4 = 8   3 x 4 = 12   4 x 4 = 16   5 x 4 = 20   10 x 4 = 40

How many altogether?

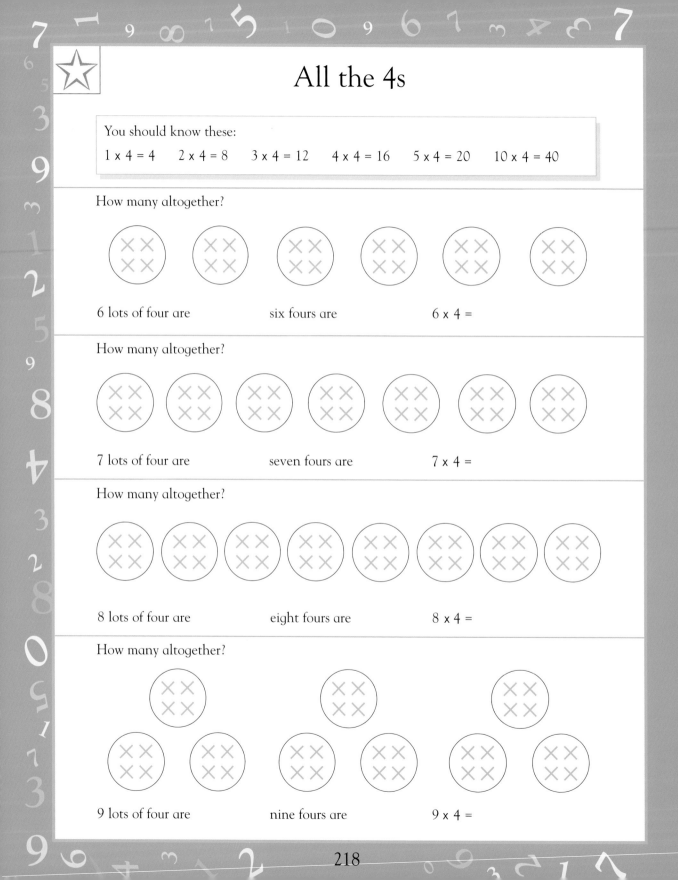

6 lots of four are   six fours are   6 x 4 =

How many altogether?

7 lots of four are   seven fours are   7 x 4 =

How many altogether?

8 lots of four are   eight fours are   8 x 4 =

How many altogether?

9 lots of four are   nine fours are   9 x 4 =

All the 4s again

Cover the 4 times table with a piece of paper so you can't see the numbers.
Write the answers. Be as fast as you can, but get them right!

2 x 4 =	5 x 4 =	6 x 4 =
3 x 4 =	7 x 4 =	4 x 12 =
4 x 4 =	4 x 11 =	4 x 1 =
5 x 4 =	3 x 4 =	5 x 4 =
6 x 4 =	6 x 4 =	4 x 7 =
7 x 4 =	8 x 4 =	3 x 4 =
8 x 4 =	12 x 4 =	2 x 4 =
9 x 4 =	1 x 4 =	4 x 11 =
10 x 4 =	4 x 4 =	4 x 3 =
11 x 4 =	2 x 4 =	4 x 6 =
4 x 2 =	4 x 5 =	4 x 5 =
4 x 3 =	4 x 7 =	4 x 8 =
4 x 4 =	4 x 9 =	7 x 4 =
4 x 5 =	4 x 4 =	4 x 2 =
4 x 6 =	4 x 6 =	4 x 12 =
4 x 7 =	11 x 4 =	8 x 4 =
4 x 8 =	4 x 10 =	4 x 11 =
4 x 9 =	4 x 12 =	1 x 4 =
4 x 10 =	4 x 0 =	4 x 4 =
4 x 12 =	4 x 2 =	4 x 9 =

Speed trials

You should know all of the 2, 3, 4, 5, and 10 times tables by now, but how quickly can you remember them?
Ask someone to time you as you do this page.
Remember, you must be fast but also correct!

4 x 2 =	6 x 3 =	9 x 5 =
8 x 3 =	3 x 4 =	8 x 10 =
7 x 4 =	7 x 5 =	11 x 2 =
6 x 5 =	3 x 10 =	6 x 3 =
8 x 10 =	12 x 2 =	12 x 4 =
8 x 2 =	7 x 3 =	4 x 5 =
5 x 3 =	4 x 4 =	3 x 10 =
9 x 4 =	11 x 5 =	2 x 2 =
5 x 5 =	4 x 10 =	1 x 3 =
7 x 10 =	6 x 2 =	0 x 4 =
0 x 2 =	5 x 12 =	11 x 5 =
11 x 3 =	8 x 4 =	9 x 2 =
6 x 4 =	0 x 5 =	8 x 3 =
3 x 5 =	2 x 10 =	7 x 4 =
4 x 10 =	7 x 2 =	6 x 5 =
7 x 2 =	8 x 3 =	5 x 10 =
3 x 3 =	9 x 4 =	4 x 0 =
2 x 4 =	5 x 5 =	3 x 2 =
7 x 5 =	12 x 10 =	2 x 8 =
9 x 10 =	5 x 2 =	1 x 9 =

Some of the 6s

You should already know some of the 6 times table because they are part of the 2, 3, 4, 5, and 10 times tables.

   1 x 6 = 6   2 x 6 =12   3 x 6 =18

   4 x 6 =24   5 x 6 =30   10 x 6 =60

Find out if you can remember them quickly and correctly.

Cover the 6 times table with some paper so you can't see the numbers.
Write the answers as quickly as you can.

What are three sixes?   What are ten sixes?

What are two sixes?   What are four sixes?

What is one six?   What are five sixes?

Write the answers as quickly as you can.

How many sixes are the same as 12?   How many sixes are the same as 6?

How many sixes are the same as 30?   How many sixes are the same as 18?

How many sixes are the same as 24?   How many sixes are the same as 60?

Write the answers as quickly as you can.

Multiply six by three.   Multiply six by ten.

Multiply six by two.   Multiply six by five.

Multiply six by one.   Multiply six by four.

Write the answers as quickly as you can.

4 x 6 =   2 x 6 =   10 x 6 =

5 x 6 =   1 x 6 =   3 x 6 =

Write the answers as quickly as you can.
A box contains six eggs. A man buys five boxes. How many eggs does he have?

A packet contains six sticks of gum.
How many sticks will there be in 10 packets?

The rest of the 6s

You need to learn these:

6 x 6 = 36   7 x 6 = 42   8 x 6 = 48   9 x 6 = 54   11 x 6 = 66   12 x 6 = 72

This work will help you remember the 6 times table.

Complete these sequences.

6   12   18   24   30

5 x 6 = 30   so   6 x 6 = 30 plus another 6 =

18   24   30

6 x 6 = 36   so   7 x 6 = 36 plus another 6 =

6   12   18   48   60

7 x 6 = 42   so   8 x 6 = 42 plus another 6 =

6   18   24   30

8 x 6 = 48   so   9 x 6 = 48 plus another 6 =

24   42   60

Test yourself on the rest of the 6 times table.
Cover the above part of the page with a piece of paper.

What are six sixes?   What are seven sixes?

What are twelve sixes?   What are eleven sixes?

12 x 6 =   7 x 6 =   6 x 6 =   11 x 6 =

Practise the 6s

You should know all of the 6 times table now, but how quickly can you remember it?
Ask someone to time you as you do this page.
Remember, you must be fast but also correct!

1 x 6 =	6 x 10 =	11 x 6 =
2 x 6 =	12 x 6 =	3 x 6 =
3 x 6 =	4 x 6 =	9 x 6 =
4 x 6 =	6 x 6 =	6 x 4 =
5 x 6 =	8 x 6 =	1 x 6 =
6 x 6 =	11 x 6 =	6 x 2 =
7 x 6 =	3 x 6 =	6 x 8 =
8 x 6 =	5 x 6 =	0 x 6 =
9 x 6 =	7 x 6 =	6 x 3 =
10 x 6 =	9 x 6 =	12 x 6 =
11 x 6 =	6 x 3 =	6 x 7 =
12 x 6 =	6 x 5 =	2 x 6 =
6 x 2 =	6 x 7 =	6 x 11 =
6 x 3 =	6 x 9 =	4 x 6 =
6 x 4 =	6 x 12 =	8 x 6 =
6 x 5 =	6 x 4 =	10 x 6 =
6 x 6 =	6 x 6 =	6 x 5 =
6 x 7 =	6 x 8 =	6 x 0 =
6 x 8 =	6 x 10 =	6 x 1 =
6 x 9 =	6 x 0 =	11 x 6 =

Speed trials

You should know all of the 2, 3, 4, 5, 6, and 10 times tables by now,
but how quickly can you remember them?
Ask someone to time you as you do this page.
Remember, you must be fast but also correct!

4 x 6 =	6 x 3 =	9 x 6 =
5 x 3 =	8 x 6 =	8 x 6 =
7 x 3 =	6 x 6 =	7 x 3 =
6 x 5 =	3 x 12 =	11 x 2 =
6 x 11 =	6 x 2 =	5 x 4 =
8 x 2 =	7 x 3 =	4 x 6 =
5 x 3 =	4 x 6 =	3 x 6 =
9 x 6 =	6 x 5 =	2 x 6 =
5 x 5 =	6 x 10 =	6 x 3 =
7 x 6 =	6 x 2 =	0 x 6 =
0 x 2 =	5 x 3 =	11 x 5 =
6 x 3 =	8 x 4 =	6 x 2 =
6 x 6 =	0 x 6 =	8 x 3 =
3 x 5 =	5 x 10 =	7 x 6 =
4 x 11 =	7 x 6 =	6 x 5 =
7 x 10 =	8 x 3 =	12 x 6 =
3 x 6 =	9 x 6 =	6 x 0 =
2 x 4 =	5 x 12 =	3 x 11 =
6 x 9 =	7 x 10 =	2 x 8 =
9 x 10 =	5 x 6 =	12 x 2 =

Some of the 7s

You should already know some of the 7 times table because it is part of
the 2, 3, 4, 5, 6, and 10 times tables.

1 x 7 = 7   2 x 7 =14   3 x 7 =21   4 x 7 = 28
5 x 7 =35   6 x 7 = 42   10 x 7=70

Find out if you can remember them quickly and correctly.

Cover the 7 times table with some paper and write the answers to these questions as
quickly as you can.

What are three sevens?   What are ten sevens?

What are two sevens?   What are four sevens?

What are six sevens?   What are five sevens?

Write the answers as quickly as you can.

How many sevens are the same as 14?   How many sevens are the same as 42?

How many sevens are the same as 35?   How many sevens are the same as 21?

How many sevens are the same as 28?   How many sevens are the same as 70?

Write the answers as quickly as you can.

Multiply seven by three.   Multiply seven by ten.

Multiply seven by two.   Multiply seven by five.

Multiply seven by six.   Multiply seven by four.

Write the answers as quickly as you can.

4 x 7 =   2 x 7 =   10 x 7 =

5 x 7 =   1 x 7 =   3 x 7 =

Write the answers as quickly as you can.

A bag has seven sweets. Ann buys five bags. How many sweets does she have?

How many days are there in six weeks?

The rest of the 7s

You should now know all of the 2, 3, 4, 5, 6, and 10 times tables.

You only need to learn these parts of the 7 times table.

7 x 7 = 49   8 x 7 = 56   9 x 7 = 63   11 x 7 = 77   12 x 7 = 84

This work will help you remember the 7 times table.

Complete these sequences.

7   14   21   28   35   42

6 x 7 = 42   so   7 x 7 = 42 plus another 7 =

21   28   35

7 x 7 = 49   so   8 x 7 = 49 plus another 7 =

7   14   21   56   70

8 x 7 = 56   so   9 x 7 = 56 plus another 7 =

7   21   28   35

Test yourself on the rest of the 7 times table.
Cover the section above with a piece of paper.

What are seven sevens?   What are eight sevens?

What are twelve sevens?   What are eleven sevens?

8 x 7 =   7 x 7 =   12 x 7 =   11 x 7 =

How many days are there in eight weeks?

A packet contains seven felt-tips.
How many felt-tips will there be in nine packets?

How many sevens make 56?

Practise the 7s

You should know all of the 7 times table now, but how quickly can you remember it?
Ask someone to time you as you do this page.
Remember, you must be fast but also correct!

1 x 7 =	7 x 10 =	7 x 6 =
2 x 7 =	2 x 7 =	3 x 7 =
3 x 7 =	4 x 7 =	9 x 7 =
4 x 7 =	6 x 7 =	7 x 4 =
5 x 7 =	8 x 7 =	1 x 7 =
6 x 7 =	1 x 7 =	7 x 2 =
7 x 7 =	3 x 7 =	7 x 8 =
8 x 7 =	5 x 7 =	0 x 7 =
9 x 7 =	11 x 7 =	7 x 11 =
10 x 7 =	9 x 7 =	5 x 7 =
11 x 7 =	7 x 3 =	7 x 7 =
12 x 7 =	7 x 5 =	2 x 7 =
7 x 2 =	7 x 7 =	7 x 9 =
7 x 3 =	7 x 9 =	4 x 7 =
7 x 4 =	7 x 12 =	8 x 7 =
7 x 5 =	7 x 4 =	10 x 7 =
7 x 6 =	7 x 6 =	7 x 5 =
7 x 7 =	7 x 8 =	7 x 0 =
7 x 8 =	7 x 11 =	7 x 12 =
7 x 9 =	7 x 0 =	6 x 7 =

Speed trials

You should know all of the 2, 3, 4, 5, 6, 7, and 10 times tables by now,
but how quickly can you remember them?
Ask someone to time you as you do this page.
Remember, you must be fast but also correct!

4 x 7 =	7 x 3 =	9 x 7 =
5 x 10 =	8 x 7 =	7 x 6 =
7 x 5 =	6 x 6 =	8 x 3 =
6 x 5 =	5 x 12 =	6 x 6 =
6 x 11 =	6 x 3 =	7 x 4 =
8 x 7 =	7 x 5 =	4 x 6 =
5 x 8 =	4 x 6 =	3 x 7 =
9 x 6 =	6 x 5 =	2 x 8 =
5 x 7 =	7 x 11 =	7 x 3 =
7 x 6 =	6 x 7 =	0 x 6 =
0 x 5 =	5 x 7 =	11 x 4 =
6 x 3 =	8 x 4 =	6 x 2 =
6 x 7 =	0 x 7 =	8 x 7 =
3 x 5 =	5 x 8 =	7 x 7 =
4 x 7 =	7 x 6 =	6 x 5 =
7 x 12 =	8 x 3 =	5 x 11 =
7 x 8 =	9 x 6 =	7 x 0 =
2 x 7 =	7 x 7 =	3 x 12 =
4 x 9 =	2 x 11 =	2 x 7 =
9 x 10 =	5 x 6 =	7 x 8 =

Some of the 8s

You should already know some of the 8 times table because it is part of the 2, 3, 4, 5, 6, 7, and 10 times tables.

1 x 8 =   8   2 x 8 = 16   3 x 8 = 24   4 x 8 = 32
5 x 8 = 40   6 x 8 = 48   7 x 8 = 56   10 x 8 = 80

Find out if you can remember them quickly and correctly.

Cover the 8 times table with some paper so you can't see the numbers.
Write the answers as quickly as you can.

What are three eights?   What are ten eights?

What are two eights?   What are four eights?

What are six eights?   What are five eights?

Write the answers as quickly as you can.

How many eights are the same as 16?   How many eights are the same as 40?

How many eights are the same as 32?   How many eights are the same as 24?

How many eights are the same as 56?   How many eights are the same as 48?

Write the answers as quickly as you can.

Multiply eight by three.   Multiply eight by ten.

Multiply eight by two.   Multiply eight by five.

Multiply eight by six.   Multiply eight by four.

Write the answers as quickly as you can.

6 x 8 =   2 x 8 =   10 x 8 =

5 x 8 =   7 x 8 =   3 x 8 =

Write the answers as quickly as you can.
A pizza has eight pieces. John buys six pizzas.
How many pieces does he have?

Which number multiplied by 8 gives the answer 56?

The rest of the 8s

You only need to learn these parts of the 8 times table.

8 x 8 = 64   9 x 8 = 72   11 x 8 = 88   12 x 8 = 96

This work will help you remember the 8 times table.

Complete these sequences.

8   16   24   32   40   48

7 x 8 = 56   so   8 x 8 = 56 plus another 8 =

24   32   40

8 x 8 = 64   so   8 x 8 = 64 plus another 8 =

8   16   24   64   80

8   24   40

Test yourself on the rest of the 8 times table.
Cover the section above with a piece of paper.

What are seven eights?   What are eleven eights?

What are twelve eights?   What are nine eights?

11 x 8 =   12 x 8 =   9 x 8 =   10 x 8 =

What number multiplied by 8 gives the answer 72?

A number multiplied by 8 gives the answer 80. What is the number?

David puts out building bricks in piles of 8.
How many bricks will there be in 10 piles?

What number multiplied by 5 gives the answer 40?

How many 8s make 72?

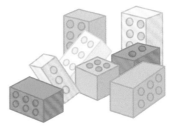

Practise the 8s

You should know all of the 8 times table now, but how quickly can you remember it?
Ask someone to time you as you do this page.
Remember, you must be fast but also correct!

1 x 8 =	8 x 10 =	8 x 6 =
2 x 8 =	2 x 8 =	3 x 8 =
3 x 8 =	4 x 8 =	9 x 8 =
4 x 8 =	6 x 8 =	8 x 4 =
5 x 8 =	8 x 8 =	11 x 8 =
6 x 8 =	12 x 8 =	8 x 2 =
7 x 8 =	1 x 8 =	7 x 8 =
8 x 8 =	3 x 8 =	12 x 8 =
9 x 8 =	5 x 8 =	8 x 3 =
10 x 8 =	7 x 8 =	5 x 8 =
11 x 8 =	8 x 3 =	8 x 8 =
12 x 8 =	8 x 5 =	2 x 8 =
8 x 2 =	8 x 8 =	8 x 9 =
8 x 3 =	8 x 9 =	4 x 8 =
8 x 4 =	8 x 11 =	8 x 7 =
8 x 5 =	8 x 4 =	10 x 8 =
8 x 6 =	8 x 6 =	8 x 12 =
8 x 7 =	8 x 8 =	8 x 0 =
8 x 8 =	8 x 10 =	8 x 11 =
8 x 9 =	8 x 0 =	12 x 8 =

Speed trials

You should know all of the 2, 3, 4, 5, 6, 7, 8, and 10 times tables now,
but how quickly can you remember them?
Ask someone to time you as you do this page.
Remember, you must be fast but also correct!

4 x 8 =	7 x 8 =	9 x 8 =
5 x 11 =	8 x 7 =	7 x 6 =
7 x 8 =	6 x 8 =	8 x 3 =
8 x 5 =	8 x 11 =	8 x 8 =
6 x 11 =	6 x 3 =	7 x 4 =
8 x 7 =	7 x 7 =	0 x 8 =
5 x 8 =	5 x 6 =	3 x 7 =
9 x 8 =	6 x 7 =	2 x 8 =
8 x 8 =	7 x 12 =	7 x 3 =
7 x 6 =	6 x 9 =	0 x 8 =
7 x 5 =	5 x 8 =	12 x 8 =
6 x 8 =	8 x 4 =	6 x 2 =
6 x 7 =	0 x 8 =	8 x 6 =
5 x 7 =	5 x 9 =	7 x 8 =
8 x 4 =	7 x 6 =	6 x 5 =
7 x 11 =	8 x 3 =	8 x 10 =
2 x 8 =	9 x 6 =	8 x 7 =
4 x 7 =	4 x 12 =	5 x 12 =
6 x 9 =	9 x 10 =	8 x 2 =
9 x 10 =	6 x 6 =	8 x 9 =

Some of the 9s

You should already know nearly all of the 9 times table because it is part of the 2, 3, 4, 5, 6, 7, 8, and 10 times tables.

1 x 9 = 9	2 x 9 = 18	3 x 9 = 27	4 x 9 = 36	5 x 9 = 45
6 x 9 = 54	7 x 9 = 63	8 x 9 = 72	10 x 9 = 90	

Find out if you can remember them quickly and correctly.

Cover the 9 times table with some paper so you can't see the numbers.
Write the answers as quickly as you can.

What are three nines?   What are ten nines?

What are two nines?   What are four nines?

What are six nines?   What are five nines?

What are seven nines?   What are eight nines?

Write the answers as quickly as you can.
How many nines are the same as 18?   How many nines are the same as 54?

How many nines are the same as 90?   How many nines are the same as 27?

How many nines are the same as 72?   How many nines are the same as 36?

How many nines are the same as 45?   How many nines are the same as 63?

Write the answers as quickly as you can.
Multiply nine by seven.   Multiply nine by ten.

Multiply nine by two.   Multiply nine by five.

Multiply nine by six.   Multiply nine by four.

Multiply nine by three.   Multiply nine by eight.

Write the answers as quickly as you can.

6 x 9 =	2 x 9 =	10 x 9 =
5 x 9 =	3 x 9 =	8 x 9 =
0 x 9 =	7 x 9 =	4 x 9 =

The rest of the 9s

You only need to learn these parts of the 9 times table.

$9 \times 9 = 81$ $9 \times 11 = 99$ $9 \times 12 = 108$

This work will help you remember the 9 times table.

Complete these sequences.

9   18   27   36   45   54

$8 \times 9 = 72$ so $9 \times 9 = 72$ plus another 9 =

45   54   63

9   18   27   72   90

9   27   45

Look for patterns in the 9 times table up to 10 x 9.

1	x	9	=	09
2	x	9	=	18
3	x	9	=	27
4	x	9	=	36
5	x	9	=	45
6	x	9	=	54
7	x	9	=	63
8	x	9	=	72
9	x	9	=	81
10	x	9	=	90

Write down any patterns you can see. There is more than one!

Practise the 9s

You should know all of the 9 times table now, but how quickly can you remember it?
Ask someone to time you as you do this page.
Remember, you must be fast but also correct!

1 x 9 =	9 x 10 =	9 x 6 =
2 x 9 =	2 x 9 =	3 x 9 =
3 x 9 =	4 x 9 =	9 x 9 =
4 x 9 =	6 x 9 =	9 x 4 =
5 x 9 =	9 x 7 =	11 x 9 =
6 x 9 =	12 x 9 =	9 x 2 =
7 x 9 =	1 x 9 =	7 x 9 =
8 x 9 =	3 x 9 =	12 x 9 =
9 x 9 =	5 x 9 =	9 x 3 =
10 x 9 =	7 x 9 =	5 x 9 =
11 x 9 =	9 x 9 =	9 x 9 =
12 x 9 =	9 x 11 =	2 x 9 =
9 x 2 =	9 x 5 =	8 x 9 =
9 x 3 =	0 x 9 =	4 x 9 =
9 x 4 =	9 x 1 =	9 x 7 =
9 x 5 =	9 x 2 =	10 x 9 =
9 x 6 =	9 x 4 =	9 x 5 =
9 x 7 =	9 x 6 =	9 x 0 =
9 x 8 =	9 x 8 =	9 x 11 =
9 x 9 =	9 x 12 =	12 x 9 =

Speed trials

You should know all of the times tables by now, but how quickly can you remember them?
Ask someone to time you as you do this page.
Remember, you must be fast but also correct!

6 x 8 =	4 x 8 =	8 x 12 =
9 x 12 =	9 x 8 =	7 x 9 =
5 x 8 =	6 x 6 =	8 x 5 =
7 x 5 =	8 x 9 =	8 x 7 =
6 x 4 =	6 x 4 =	7 x 4 =
8 x 8 =	7 x 3 =	4 x 9 =
5 x 11 =	5 x 9 =	6 x 7 =
9 x 8 =	6 x 8 =	4 x 6 =
8 x 3 =	7 x 7 =	7 x 8 =
7 x 7 =	6 x 9 =	6 x 9 =
9 x 5 =	7 x 8 =	11 x 8 =
4 x 8 =	8 x 4 =	6 x 5 =
6 x 7 =	0 x 9 =	8 x 8 =
2 x 9 =	10 x 12 =	7 x 6 =
8 x 4 =	7 x 6 =	6 x 8 =
7 x 12 =	8 x 7 =	9 x 10 =
2 x 8 =	9 x 6 =	8 x 4 =
4 x 7 =	8 x 6 =	7 x 11 =
6 x 9 =	11 x 9 =	5 x 8 =
9 x 9 =	6 x 7 =	8 x 9 =

Times tables for division

Knowing the times tables can also help with division sums.
Look at these examples.
3 x 6 = 18 which means that 18 ÷ 3 = 6 and that 18 ÷ 6 = 3
4 x 5 = 20 which means that 20 ÷ 4 = 5 and that 20 ÷ 5 = 4
9 x 11 = 99 which means that 99 ÷ 11 = 9 and that 99 ÷ 9 − 11

Use your knowledge of the times tables to work out these division sums.

3 x 8 = 24 which means that 24 ÷ 3 =   and that 24 ÷ 8 =

4 x 7 = 28 which means that 28 ÷ 4 =   and that 28 ÷ 7 =

3 x 5 = 15 which means that 15 ÷ 3 =   and that 15 ÷ 5 =

4 x 3 = 12 which means that 12 ÷ 3 =   and that 12 ÷ 4 =

3 x 11 = 33 which means that 33 ÷ 3 =   and that 33 ÷ 11 =

4 x 8 = 32 which means that 32 ÷ 4 =   and that 32 ÷ 8 =

3 x 9 = 27 which means that 27 ÷ 3 =   and that 27 ÷ 9 =

4 x 12 = 48 which means that 48 ÷ 4 =   and that 48 ÷ 12 =

These division sums help practise the 3 and 4 times tables.

20 ÷ 4 =	33 ÷ 3 =	16 ÷ 4 =
24 ÷ 4 =	27 ÷ 3 =	30 ÷ 3 =
12 ÷ 3 =	18 ÷ 3 =	28 ÷ 4 =
24 ÷ 3 =	48 ÷ 4 =	21 ÷ 3 =

How many fours in 36?   Divide 27 by three.

Divide 28 by 4.   How many threes in 21?

How many fives in 35?   Divide 40 by 5.

Divide 15 by 3.   How many eights in 48?

Times tables for division

This page will help you remember times tables by dividing by 2, 3, 4, 5, and 10.

20 ÷ 5 =   4   18 ÷ 3 =   6   60 ÷ 5 =   12

Complete the sums.

44 ÷ 4 =	14 ÷ 2 =	32 ÷ 4 =
25 ÷ 5 =	21 ÷ 3 =	16 ÷ 4 =
24 ÷ 4 =	28 ÷ 4 =	12 ÷ 2 =
45 ÷ 5 =	60 ÷ 5 =	12 ÷ 3 =
10 ÷ 2 =	40 ÷ 10 =	12 ÷ 4 =
20 ÷ 10 =	20 ÷ 2 =	20 ÷ 2 =
6 ÷ 2 =	18 ÷ 3 =	20 ÷ 4 =
24 ÷ 3 =	32 ÷ 4 =	20 ÷ 5 =
30 ÷ 5 =	40 ÷ 5 =	20 ÷ 10 =
36 ÷ 3 =	33 ÷ 3 =	18 ÷ 2 =
40 ÷ 5 =	6 ÷ 2 =	18 ÷ 3 =
21 ÷ 3 =	15 ÷ 3 =	15 ÷ 3 =
14 ÷ 2 =	24 ÷ 4 =	15 ÷ 5 =
27 ÷ 3 =	15 ÷ 5 =	24 ÷ 3 =
48 ÷ 4 =	10 ÷ 10 =	24 ÷ 2 =
15 ÷ 5 =	4 ÷ 2 =	50 ÷ 5 =
15 ÷ 3 =	9 ÷ 3 =	55 ÷ 5 =
20 ÷ 5 =	4 ÷ 4 =	30 ÷ 3 =
20 ÷ 4 =	10 ÷ 5 =	30 ÷ 5 =
16 ÷ 2 =	110 ÷ 10 =	30 ÷ 10 =

Times tables for division

This page will help you remember times tables by dividing by 2, 3, 4, 5, 6, 10, 11, and 12.

$30 \div 6 =$ 5 $12 \div 6 =$ 2 $66 \div 11 =$ 6

Complete the sums.

$18 \div 6 =$	$27 \div 3 =$	$48 \div 6 =$
$30 \div 10 =$	$18 \div 6 =$	$35 \div 5 =$
$14 \div 2 =$	$22 \div 2 =$	$36 \div 4 =$
$18 \div 3 =$	$24 \div 6 =$	$24 \div 3 =$
$20 \div 4 =$	$24 \div 3 =$	$20 \div 2 =$
$15 \div 5 =$	$24 \div 4 =$	$33 \div 3 =$
$36 \div 6 =$	$30 \div 10 =$	$25 \div 5 =$
$55 \div 5 =$	$18 \div 2 =$	$32 \div 4 =$
$48 \div 4 =$	$18 \div 3 =$	$24 \div 2 =$
$15 \div 3 =$	$36 \div 4 =$	$16 \div 2 =$
$16 \div 4 =$	$36 \div 6 =$	$42 \div 6 =$
$25 \div 5 =$	$40 \div 5 =$	$5 \div 5 =$
$6 \div 6 =$	$120 \div 10 =$	$4 \div 4 =$
$10 \div 10 =$	$16 \div 4 =$	$28 \div 4 =$
$42 \div 6 =$	$12 \div 6 =$	$14 \div 2 =$
$24 \div 4 =$	$48 \div 12 =$	$24 \div 6 =$
$54 \div 6 =$	$54 \div 6 =$	$18 \div 6 =$
$99 \div 11 =$	$60 \div 6 =$	$54 \div 6 =$
$30 \div 6 =$	$66 \div 6 =$	$60 \div 6 =$
$30 \div 5 =$	$30 \div 6 =$	$40 \div 5 =$

Times tables for division

This page will help you remember times tables by dividing by 2, 3, 4, 5, 6, and 7.

$14 \div 7 = \quad 2$ $\qquad 28 \div 7 = \quad 4$ $\qquad 84 \div 7 = \quad 12$

Complete the sums.

$21 \div 7 =$	$77 \div 7 =$	$84 \div 7 =$
$35 \div 5 =$	$28 \div 7 =$	$35 \div 5 =$
$14 \div 2 =$	$24 \div 6 =$	$35 \div 7 =$
$18 \div 6 =$	$24 \div 4 =$	$24 \div 6 =$
$20 \div 5 =$	$24 \div 2 =$	$21 \div 3 =$
$15 \div 3 =$	$21 \div 7 =$	$70 \div 7 =$
$36 \div 4 =$	$42 \div 7 =$	$42 \div 7 =$
$55 \div 5 =$	$18 \div 3 =$	$32 \div 4 =$
$18 \div 2 =$	$49 \div 7 =$	$27 \div 3 =$
$15 \div 5 =$	$36 \div 4 =$	$16 \div 4 =$
$48 \div 4 =$	$36 \div 3 =$	$42 \div 6 =$
$25 \div 5 =$	$40 \div 5 =$	$45 \div 5 =$
$7 \div 7 =$	$70 \div 7 =$	$84 \div 7 =$
$63 \div 7 =$	$24 \div 3 =$	$24 \div 3 =$
$42 \div 7 =$	$42 \div 6 =$	$14 \div 7 =$
$24 \div 2 =$	$48 \div 6 =$	$24 \div 4 =$
$54 \div 6 =$	$54 \div 6 =$	$18 \div 3 =$
$28 \div 7 =$	$60 \div 6 =$	$56 \div 7 =$
$30 \div 6 =$	$66 \div 6 =$	$63 \div 7 =$
$35 \div 7 =$	$25 \div 5 =$	$48 \div 6 =$

Times tables for division

This page will help you remember times tables by dividing by
2, 3, 4, 5, 6, 7, 8, and 9.

$16 \div 8 =$ 2 $35 \div 7 =$ 5 $27 \div 9 =$ 3

$42 \div 6 =$	$81 \div 9 =$	$56 \div 7 =$
$32 \div 8 =$	$56 \div 7 =$	$45 \div 5 =$
$14 \div 7 =$	$63 \div 7 =$	$35 \div 7 =$
$48 \div 4 =$	$24 \div 6 =$	$18 \div 9 =$
$63 \div 7 =$	$22 \div 2 =$	$21 \div 3 =$
$72 \div 9 =$	$72 \div 9 =$	$28 \div 7 =$
$72 \div 8 =$	$42 \div 6 =$	$60 \div 5 =$
$56 \div 7 =$	$108 \div 9 =$	$32 \div 8 =$
$18 \div 6 =$	$14 \div 7 =$	$27 \div 9 =$
$81 \div 9 =$	$36 \div 4 =$	$16 \div 8 =$
$63 \div 9 =$	$36 \div 6 =$	$72 \div 6 =$
$45 \div 5 =$	$48 \div 8 =$	$45 \div 9 =$
$54 \div 9 =$	$21 \div 7 =$	$40 \div 4 =$
$70 \div 7 =$	$24 \div 3 =$	$24 \div 8 =$
$42 \div 7 =$	$40 \div 8 =$	$63 \div 7 =$
$30 \div 5 =$	$45 \div 9 =$	$24 \div 6 =$
$54 \div 6 =$	$54 \div 6 =$	$18 \div 6 =$
$56 \div 8 =$	$99 \div 9 =$	$96 \div 8 =$
$66 \div 6 =$	$63 \div 9 =$	$99 \div 9 =$
$35 \div 7 =$	$50 \div 5 =$	$48 \div 8 =$

Times tables practice grids

This is a times tables grid.

X	3	4	5
7	21	28	35
8	24	32	40

Complete each times tables grid.

X	1	3	5	7	9
2					
3					

X	4	6
6		
7		
8		

X	6	7	8	9	11
3					
4					
5					

X	10	7	8	4
3				
5				
7				

X	6	2	4	12
5				
10				

X	8	7	9	6
9				
7				

Times tables practice grids

Here are some more times tables grids.

X	2	4	6
5			
7			

X	11	3	9	2
5				
6				
7				

X	2	3	4	5
8				
9				

X	10	9	8	7
6				
5				
4				

X	3	12
2		
3		
4		
5		
6		
7		

X	2	4	6	8
1				
3				
5				
7				
9				
0				

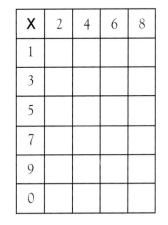

Times tables practise grids

Here are some more times tables grids.

X	8	9
7		
8		

X	9	8	7	6	5	4
9						
8						
7						

X	2	5	9
4			
7			
8			

X	2	3	4	5	7
4					
6					
8					

X	3	5	12
2			
8			
6			
0			
4			
7			

X	8	7	11	6
7				
9				
0				
10				
8				
6				

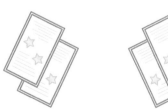

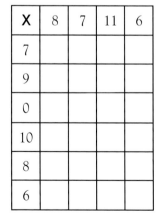

Speed trials

Try this final test.

27 ÷ 3 =	4 x 9 =	14 ÷ 2 =
7 x 9 =	18 ÷ 2 =	9 x 9 =
64 ÷ 8 =	6 x 8 =	15 ÷ 3 =
90 ÷ 10 =	21 ÷ 3 =	8 x 12 =
6 x 8 =	9 x 7 =	24 ÷ 3 =
45 ÷ 9 =	32 ÷ 4 =	7 x 8 =
3 x 12 =	4 x 11 =	30 ÷ 5 =
9 x 5 =	45 ÷ 5 =	6 x 6 =
48 ÷ 6 =	8 x 5 =	42 ÷ 6 =
7 x 7 =	42 ÷ 6 =	9 x 12 =
3 x 11 =	7 x 4 =	49 ÷ 7 =
56 ÷ 8 =	35 ÷ 7 =	8 x 6 =
36 ÷ 4 =	9 x 3 =	72 ÷ 8 =
24 ÷ 3 =	24 ÷ 8 =	9 x 7 =
36 ÷ 9 =	8 x 2 =	54 ÷ 9 =
6 x 7 =	36 ÷ 9 =	7 x 6 =
4 x 4 =	6 x 12 =	10 ÷ 10 =
32 ÷ 8 =	80 ÷ 10 =	7 x 11 =
49 ÷ 7 =	11 x 9 =	16 ÷ 8 =
25 ÷ 5 =	16 ÷ 2 =	7 x 9 =
56 ÷ 7 =	54 ÷ 9 =	63 ÷ 7 =

Answer Section

More Times Tables Practice

This 8-page section provides answers to all the activities in this chapter. This will enable you to mark your child's work or can be used by them if they prefer to do their own marking.

Speed trials

Write the answers as fast as you can, but get them right!

4 x 10 = 40	8 x 2 = 16	6 x 5 = 30

Write the answers as fast as you can, but get them right!

3 x 2 = 6	0 x 5 = 0	3 x 10 = 30	0 x 3 = 0
5 x 2 = 10	10 x 5 = 50	5 x 10 = 50	10 x 3 = 30
1 x 2 = 2	8 x 5 = 40	1 x 10 = 10	8 x 3 = 24
4 x 2 = 8	6 x 5 = 30	4 x 10 = 40	6 x 3 = 18
12 x 2 = 24	2 x 5 = 10	7 x 10 = 70	2 x 3 = 6
2 x 2 = 4	7 x 5 = 35	2 x 10 = 20	7 x 3 = 21
6 x 2 = 12	4 x 5 = 20	6 x 10 = 60	4 x 3 = 12
8 x 2 = 16	1 x 5 = 5	8 x 10 = 80	1 x 3 = 3
10 x 2 = 20	5 x 5 = 25	10 x 11 = 110	5 x 3 = 15
0 x 2 = 0	3 x 5 = 15	0 x 10 = 0	3 x 3 = 9
9 x 2 = 18	5 x 3 = 15	9 x 10 = 90	6 x 4 = 24
2 x 7 = 14	5 x 8 = 40	10 x 7 = 70	3 x 4 = 12
2 x 11 = 22	5 x 6 = 30	10 x 1 = 10	7 x 12 = 84
2 x 4 = 8	5 x 9 = 45	10 x 4 = 40	4 x 4 = 16
3 x 7 = 21	5 x 7 = 35	10 x 7 = 70	10 x 4 = 40
2 x 5 = 10	11 x 4 = 44	10 x 12 = 120	8 x 4 = 32
2 x 9 = 18	5 x 1 = 5	10 x 9 = 90	0 x 4 = 0
2 x 6 = 12	4 x 7 = 28	10 x 6 = 60	9 x 4 = 36
2 x 8 = 16	5 x 11 = 55	10 x 8 = 80	5 x 4 = 20
12 x 3 = 36	5 x 2 = 10	10 x 3 = 30	2 x 4 = 8

All the 3s

You will need to know these:

1 x 3 = 3   2 x 3 = 6   3 x 3 = 9   4 x 3 = 12   5 x 3 = 15   10 x 3 = 30

How many altogether?

6 lots of three are 18   six threes are 18   6 x 3 = 18

How many altogether?

7 lots of three are 21   seven threes are 21   7 x 3 = 21

How many altogether?

8 lots of three are 24   eight threes are 24   8 x 3 = 24

How many altogether?

9 lots of three are 27   nine threes are 27   9 x 3 = 27

All the 3s again

You should know all of the 3 times table by now.

1 x 3 = 3   2 x 3 = 6   3 x 3 = 9   4 x 3 = 12   5 x 3 = 15   6 x 3 = 18
7 x 3 = 21   8 x 3 = 24   9 x 3 = 27   10 x 3 = 30   11 x 3 = 33   12 x 3 = 36

Say these through to yourself a few times.

Cover the 3 times table with a piece of paper so you can't see the numbers. Write the answers. Be as fast as you can, but get them right!

2 x 3 = 6	5 x 3 = 15	6 x 3 = 18
3 x 3 = 9	7 x 3 = 21	9 x 3 = 27
4 x 3 = 12	9 x 3 = 27	4 x 3 = 12
5 x 3 = 15	4 x 3 = 12	5 x 3 = 15
6 x 3 = 18	6 x 3 = 18	3 x 7 = 21
7 x 3 = 21	8 x 3 = 24	3 x 4 = 12
8 x 3 = 24	10 x 3 = 30	2 x 3 = 6
9 x 3 = 27	11 x 3 = 33	12 x 3 = 36
10 x 3 = 30	12 x 3 = 36	3 x 9 = 27
11 x 3 = 33	2 x 3 = 6	3 x 6 = 18
3 x 2 = 6	3 x 5 = 15	3 x 5 = 15
3 x 3 = 9	3 x 7 = 21	3 x 8 = 24
3 x 4 = 12	3 x 9 = 27	7 x 3 = 21
3 x 5 = 15	3 x 4 = 12	3 x 2 = 6
3 x 6 = 18	3 x 6 = 18	3 x 11 = 33
3 x 7 = 21	3 x 8 = 24	8 x 3 = 24
3 x 8 = 24	3 x 10 = 30	3 x 10 = 30
3 x 9 = 27	3 x 1 = 3	1 x 3 = 3
3 x 10 = 30	3 x 0 = 0	3 x 3 = 9
3 x 12 = 36	3 x 2 = 6	3 x 9 = 27

All the 4s

You should know these:

1 x 4 = 4   2 x 4 = 8   3 x 4 = 12   4 x 4 = 16   5 x 4 = 20   10 x 4 = 40

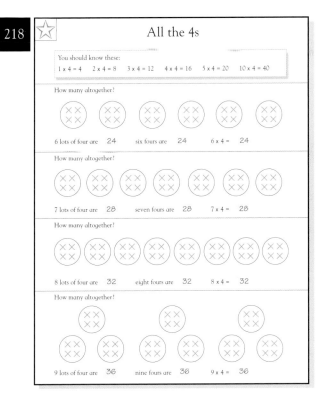

How many altogether?

6 lots of four are 24   six fours are 24   6 x 4 = 24

How many altogether?

7 lots of four are 28   seven fours are 28   7 x 4 = 28

How many altogether?

8 lots of four are 32   eight fours are 32   8 x 4 = 32

How many altogether?

9 lots of four are 36   nine fours are 36   9 x 4 = 36

All the 4s again

You should know all of the 4 times table by now.

1 x 4 = 4   2 x 4 = 8   3 x 4 = 12   4 x 4 = 16   5 x 4 = 20   6 x 4 = 24
7 x 4 = 28   8 x 4 = 32   9 x 4 = 36   10 x 4 = 40   11 x 4 = 44   12 x 4 = 48

Say these through to yourself a few times.

Cover the 4 times table with a piece of paper so you can't see the numbers.
Write the answers. Be as fast as you can, but get them right!

2 x 4 = 8	5 x 4 = 20	6 x 4 = 24
3 x 4 = 12	7 x 4 = 28	4 x 12 = 48
4 x 4 = 16	4 x 11 = 44	4 x 1 = 4
5 x 4 = 20	3 x 4 = 12	5 x 4 = 20
6 x 4 = 24	6 x 4 = 24	4 x 7 = 28
7 x 4 = 28	8 x 4 = 32	3 x 4 = 12
8 x 4 = 32	12 x 4 = 48	2 x 4 = 8
9 x 4 = 36	1 x 4 = 4	4 x 11 = 44
10 x 4 = 40	4 x 4 = 16	4 x 3 = 12
11 x 4 = 44	2 x 4 = 8	4 x 6 = 24
4 x 2 = 8	4 x 5 = 20	4 x 5 = 20
4 x 3 = 12	4 x 7 = 28	4 x 8 = 32
4 x 4 = 16	4 x 9 = 36	7 x 4 = 28
4 x 5 = 20	4 x 4 = 16	4 x 2 = 8
4 x 6 = 24	4 x 6 = 24	4 x 12 = 48
4 x 7 = 28	11 x 4 = 44	8 x 4 = 32
4 x 8 = 32	4 x 10 = 40	4 x 11 = 44
4 x 9 = 36	4 x 12 = 48	1 x 4 = 4
4 x 10 = 40	4 x 0 = 0	4 x 4 = 16
4 x 12 = 48	4 x 2 = 8	4 x 9 = 36

Speed trials

You should know all of the 2, 3, 4, 5, and 10 times tables by now, but how quickly can you remember them?
Ask someone to time you as you do this page.
Remember, you must be fast but also correct!

4 x 2 = 8	6 x 3 = 18	9 x 5 = 45
8 x 3 = 24	3 x 4 = 12	8 x 10 = 80
7 x 4 = 28	7 x 5 = 35	11 x 2 = 22
6 x 5 = 30	3 x 10 = 30	6 x 3 = 18
8 x 10 = 80	12 x 2 = 24	12 x 4 = 48
8 x 2 = 16	7 x 3 = 21	4 x 5 = 20
5 x 3 = 15	4 x 4 = 16	3 x 10 = 30
9 x 4 = 36	11 x 5 = 55	2 x 2 = 4
5 x 5 = 25	4 x 10 = 40	1 x 3 = 3
7 x 10 = 70	6 x 2 = 12	0 x 4 = 0
0 x 2 = 0	5 x 12 = 60	11 x 5 = 55
11 x 3 = 33	8 x 4 = 32	9 x 2 = 18
6 x 4 = 24	0 x 5 = 0	8 x 3 = 24
3 x 5 = 15	2 x 10 = 20	7 x 4 = 28
4 x 10 = 40	7 x 12 = 14	6 x 5 = 30
7 x 2 = 14	8 x 3 = 24	5 x 10 = 50
3 x 3 = 9	9 x 4 = 36	4 x 0 = 0
2 x 4 = 8	5 x 5 = 25	3 x 2 = 6
7 x 5 = 35	12 x 10 = 120	2 x 8 = 16
9 x 10 = 90	5 x 2 = 10	1 x 9 = 9

Some of the 6s

You should already know some of the 6 times table because they are part of the 2, 3, 4, 5, and 10 times tables.

1 x 6 = 6   2 x 6 = 12   3 x 6 = 18
4 x 6 = 24   5 x 6 = 30   10 x 6 = 60

Find out if you can remember them quickly and correctly.

Cover the 6 times table with some paper so you can't see the numbers.
Write the answers as quickly as you can.

What are three sixes? 18   What are ten sixes? 60

What are two sixes? 12   What are four sixes? 24

What is one six? 6   What are five sixes? 30

Write the answers as quickly as you can.

How many sixes are the same as 12? 2   How many sixes are the same as 6? 1

How many sixes are the same as 30? 5   How many sixes are the same as 18? 3

How many sixes are the same as 24? 4   How many sixes are the same as 60? 10

Write the answers as quickly as you can.

Multiply six by three. 18   Multiply six by ten. 60

Multiply six by two. 12   Multiply six by five. 30

Multiply six by one. 6   Multiply six by four. 24

Write the answers as quickly as you can.

4 x 6 = 24   2 x 6 = 12   10 x 6 = 60

5 x 6 = 30   1 x 6 = 6   3 x 6 = 18

Write the answers as quickly as you can.
A box contains six eggs. A man buys five boxes. How many eggs does he have? 30

A packet contains six sticks of gum.
How many sticks will there be in 10 packets? 60

The rest of the 6s

You need to learn these:
6 x 6 = 36   7 x 6 = 42   8 x 6 = 48   9 x 6 = 54   11 x 6 = 66   12 x 6 = 72

This work will help you remember the 6 times table.

Complete these sequences.

6   12   18   24   30   **36**   42   **48**   54   **60**

5 x 6 = 30   so   6 x 6 = 30 plus another 6 =   **36**

18   24   30   **36**   **42**   **48**   54   **60**

6 x 6 = 36   so   7 x 6 = 36 plus another 6 =   **42**

6   12   18   **24**   **30**   **36**   42   48   **54**   60

7 x 6 = 42   so   8 x 6 = 42 plus another 6 =   **48**

6   12   18   **24**   30   **36**   42   48   **54**   60

8 x 6 = 48   so   9 x 6 = 48 plus another 6 =   **54**

6   **12**   **18**   24   30   36   42   **48**   **54**   60

Test yourself on the rest of the 6 times table.
Cover the above part of the page with a piece of paper.

What are six sixes?   **36**   What are seven sixes?   **42**

What are twelve sixes?   **72**   What are eleven sixes?   **66**

12 x 6 =   **72**   7 x 6 =   **42**   6 x 6 =   **36**   11 x 6 =   **66**

Practise the 6s

You should know all of the 6 times table now, but how quickly can you remember it?
Ask someone to time you as you do this page.
Remember, you must be fast but also correct!

1 x 6 = **6**	6 x 10 = **60**	11 x 6 = **66**
2 x 6 = **12**	12 x 6 = **72**	3 x 6 = **18**
3 x 6 = **18**	4 x 6 = **24**	9 x 6 = **54**
4 x 6 = **24**	6 x 6 = **36**	6 x 4 = **24**
5 x 6 = **30**	8 x 6 = **48**	1 x 6 = **6**
6 x 6 = **36**	11 x 6 = **66**	6 x 2 = **12**
7 x 6 = **42**	3 x 6 = **18**	6 x 8 = **48**
8 x 6 = **48**	5 x 6 = **30**	0 x 6 = **0**
9 x 6 = **54**	7 x 6 = **42**	6 x 3 = **18**
10 x 6 = **60**	9 x 6 = **54**	12 x 6 = **72**
11 x 6 = **66**	6 x 3 = **18**	6 x 7 = **42**
12 x 6 = **72**	6 x 5 = **30**	2 x 6 = **12**
6 x 2 = **12**	6 x 7 = **42**	6 x 11 = **66**
6 x 3 = **18**	6 x 9 = **54**	4 x 6 = **24**
6 x 4 = **24**	6 x 12 = **72**	8 x 6 = **48**
6 x 5 = **30**	6 x 4 = **24**	10 x 6 = **60**
6 x 6 = **36**	6 x 6 = **36**	6 x 5 = **30**
6 x 7 = **42**	6 x 8 = **48**	6 x 0 = **0**
6 x 8 = **48**	6 x 10 = **60**	6 x 1 = **6**
6 x 9 = **54**	6 x 0 = **0**	11 x 6 = **66**

Speed trials

You should know all of the 2, 3, 4, 5, 6, and 10 times tables by now,
but how quickly can you remember them?
Ask someone to time you as you do this page.
Remember, you must be fast but also correct!

4 x 6 = **24**	6 x 3 = **18**	9 x 6 = **54**
5 x 3 = **15**	8 x 6 = **48**	8 x 6 = **48**
7 x 3 = **21**	6 x 6 = **36**	7 x 3 = **21**
6 x 5 = **30**	3 x 12 = **36**	11 x 2 = **22**
6 x 11 = **66**	6 x 2 = **12**	5 x 4 = **20**
8 x 2 = **16**	7 x 3 = **21**	4 x 6 = **24**
5 x 3 = **15**	4 x 6 = **24**	3 x 6 = **18**
9 x 6 = **54**	6 x 5 = **30**	2 x 6 = **12**
5 x 5 = **25**	6 x 10 = **60**	6 x 3 = **18**
7 x 6 = **42**	6 x 2 = **12**	0 x 6 = **0**
0 x 2 = **0**	5 x 3 = **15**	11 x 5 = **55**
6 x 3 = **18**	8 x 4 = **32**	6 x 2 = **12**
6 x 6 = **36**	0 x 6 = **0**	8 x 3 = **24**
3 x 5 = **15**	5 x 10 = **50**	7 x 6 = **42**
4 x 11 = **44**	7 x 6 = **42**	6 x 5 = **30**
7 x 10 = **70**	8 x 3 = **24**	12 x 6 = **72**
3 x 6 = **18**	9 x 6 = **54**	6 x 0 = **0**
2 x 4 = **8**	5 x 12 = **60**	3 x 11 = **33**
6 x 9 = **54**	7 x 10 = **70**	2 x 8 = **16**
9 x 10 = **90**	5 x 6 = **30**	12 x 2 = **24**

Some of the 7s

You should already know some of the 7 times table because it is part of
the 2, 3, 4, 5, 6, and 10 times tables.
1 x 7 = 7   2 x 7 = 14   3 x 7 = 21   4 x 7 = 28
5 x 7 = 35   6 x 7 = 42   10 x 7 = 70
Find out if you can remember them quickly and correctly.

Cover the 7 times table with some paper and write the answers to these questions as
quickly as you can.

What are three sevens?   **21**   What are ten sevens?   **70**

What are two sevens?   **14**   What are four sevens?   **28**

What are six sevens?   **42**   What are five sevens?   **35**

Write the answers as quickly as you can.

How many sevens are the same as 14?   **2**   How many sevens are the same as 42?   **6**

How many sevens are the same as 35?   **5**   How many sevens are the same as 21?   **3**

How many sevens are the same as 28?   **4**   How many sevens are the same as 70?   **10**

Write the answers as quickly as you can.

Multiply seven by three.   **21**   Multiply seven by ten.   **70**

Multiply seven by two.   **14**   Multiply seven by five.   **35**

Multiply seven by six.   **42**   Multiply seven by four.   **28**

Write the answers as quickly as you can.

4 x 7 =   **28**   2 x 7 =   **14**   10 x 7 =   **70**

5 x 7 =   **35**   1 x 7 =   **7**   3 x 7 =   **21**

Write the answers as quickly as you can.

A bag has seven sweets. Ann buys five bags. How many sweets does she have?   **35**

How many days are there in six weeks?   **42**

The rest of the 7s

You should now know all of the 2, 3, 4, 5, 6, and 10 times tables.
You only need to learn these parts of the 7 times table.
7 x 7 = 49   8 x 7 = 56   9 x 7 = 63   11 x 7 = 77   12 x 7 = 84

This work will help you remember the 7 times table.

Complete these sequences.

7   14   21   28   35   42   **49**   **56**   **63**   70

6 x 7 = 42   so   7 x 7 = 42 plus another 7 =   **49**

21   28   35   42   **49**   **56**   **63**   70

7 x 7 = 49   so   8 x 7 = 49 plus another 7 =   **56**

7   14   21   **28**   **35**   42   49   56   63   70

8 x 7 = 56   so   9 x 7 = 56 plus another 7 =   **63**

7   14   21   28   35   42   49   56   63   70

Test yourself on the rest of the 7 times table.
Cover the section above with a piece of paper.

What are seven sevens?   **49**   What are eight sevens?   **56**

What are twelve sevens?   **84**   What are eleven sevens?   **77**

8 x 7 =   **56**   7 x 7 =   **49**   12 x 7 =   **84**   11 x 7 =   **77**

How many days are there in eight weeks?   **56**

A packet contains seven felt-tips.
How many felt-tips will there be in nine packets?   **63**

How many sevens make 56?   **8**

Practise the 7s

You should know all of the 7 times table now, but how quickly can you remember it?
Ask someone to time you as you do this page.
Remember, you must be fast but also correct!

1 x 7 =	7	7 x 10 =	70	7 x 6 =	42
2 x 7 =	14	2 x 7 =	14	3 x 7 =	21
3 x 7 =	21	4 x 7 =	28	9 x 7 =	63
4 x 7 =	28	6 x 7 =	42	7 x 4 =	28
5 x 7 =	35	8 x 7 =	56	1 x 7 =	7
6 x 7 =	42	1 x 7 =	7	7 x 2 =	14
7 x 7 =	49	3 x 7 =	21	7 x 8 =	56
8 x 7 =	56	5 x 7 =	35	0 x 7 =	0
9 x 7 =	63	11 x 7 =	77	7 x 11 =	77
10 x 7 =	70	9 x 7 =	63	5 x 7 =	35
11 x 7 =	77	7 x 3 =	21	7 x 7 =	49
12 x 7 =	84	7 x 5 =	35	2 x 7 =	14
7 x 2 =	14	7 x 7 =	49	7 x 9 =	63
7 x 3 =	21	7 x 9 =	63	4 x 7 =	28
7 x 4 =	28	7 x 12 =	84	8 x 7 =	56
7 x 5 =	35	7 x 4 =	28	10 x 7 =	70
7 x 6 =	42	7 x 6 =	42	7 x 5 =	35
7 x 7 =	49	7 x 8 =	56	7 x 0 =	0
7 x 8 =	56	7 x 11 =	77	7 x 12 =	84
7 x 9 =	63	7 x 0 =	0	6 x 7 =	42

Speed trials

You should know all of the 2, 3, 4, 5, 6, 7, and 10 times tables by now,
but how quickly can you remember them?
Ask someone to time you as you do this page.
Remember, you must be fast but also correct!

4 x 7 =	28	7 x 3 =	21	9 x 7 =	63
5 x 10 =	50	8 x 7 =	56	7 x 6 =	42
7 x 5 =	35	6 x 6 =	36	8 x 3 =	24
6 x 5 =	30	5 x 12 =	60	6 x 6 =	36
6 x 11 =	66	6 x 3 =	18	7 x 4 =	28
8 x 7 =	56	7 x 5 =	35	4 x 6 =	24
5 x 8 =	40	4 x 6 =	24	3 x 7 =	21
9 x 6 =	54	6 x 5 =	30	2 x 8 =	16
5 x 7 =	35	7 x 11 =	77	7 x 3 =	21
7 x 6 =	42	6 x 7 =	42	0 x 6 =	0
0 x 5 =	0	5 x 7 =	35	11 x 4 =	44
6 x 3 =	18	8 x 4 =	32	6 x 2 =	12
6 x 7 =	42	0 x 7 =	0	8 x 7 =	56
3 x 5 =	15	5 x 8 =	40	7 x 7 =	49
4 x 7 =	28	7 x 6 =	42	6 x 5 =	30
7 x 12 =	84	8 x 3 =	24	5 x 11 =	55
7 x 8 =	56	9 x 6 =	54	7 x 0 =	0
2 x 7 =	14	7 x 7 =	49	3 x 12 =	36
4 x 9 =	36	2 x 11 =	22	2 x 7 =	14
9 x 10 =	90	5 x 6 =	30	7 x 8 =	56

Some of the 8s

You should already know some of the 8 times table because it is part of
the 2, 3, 4, 5, 6, 7, and 10 times tables.
1 x 8 = 8   2 x 8 = 16   3 x 8 = 24   4 x 8 = 32
5 x 8 = 40   6 x 8 = 48   7 x 8 = 56   10 x 8 = 80
Find out if you can remember them quickly and correctly.

Cover the 8 times table with some paper so you can't see the numbers.
Write the answers as quickly as you can.

What are three eights?   **24**   What are ten eights?   **80**

What are two eights?   **16**   What are four eights?   **32**

What are six eights?   **48**   What are five eights?   **40**

Write the answers as quickly as you can.

How many eights are the same as 16?   **2**   How many eights are the same as 40?   **5**

How many eights are the same as 32?   **4**   How many eights are the same as 24?   **3**

How many eights are the same as 56?   **7**   How many eights are the same as 48?   **6**

Write the answers as quickly as you can.

Multiply eight by three.   **24**   Multiply eight by ten.   **80**

Multiply eight by two.   **16**   Multiply eight by five.   **40**

Multiply eight by six.   **48**   Multiply eight by four.   **32**

Write the answers as quickly as you can.

6 x 8 =   **48**   2 x 8 =   **16**   10 x 8 =   **80**

5 x 8 =   **40**   7 x 8 =   **56**   3 x 8 =   **24**

Write the answers as quickly as you can.
A pizza has eight pieces. John buys six pizzas.
How many pieces does he have?   **48**

Which number multiplied by 8 gives the answer 56?   **7**

The rest of the 8s

You only need to learn these parts of the 8 times table.
8 x 8 = 64   9 x 8 = 72   11 x 8 = 88   12 x 8 = 96

This work will help you remember the 8 times table.

Complete these sequences.

8   16   24   32   40   48   56   64   72   80

7 x 8 = 56   so   8 x 8 = 56 plus another 8 = 64

24   32   40   48   56   64   72   80

8 x 8 = 64   so   8 x 8 = 64 plus another 8 = 72

8   16   24   32   40   48   56   64   72   80

8   16   24   32   40   48   56   64   72   80

Test yourself on the rest of the 8 times table.
Cover the section above with a piece of paper.

What are seven eights?   56   What are eleven eights?   88

What are twelve eights?   96   What are nine eights?   72

11 x 8 = 88   12 x 8 = 96   9 x 8 = 72   10 x 8 = 80

What number multiplied by 8 gives the answer 72?   9

A number multiplied by 8 gives the answer 80. What is the number?   10

David puts out building bricks in piles of 8.
How many bricks will there be in 10 piles?   80

What number multiplied by 5 gives the answer 40?   8

How many 8s make 72?   9

Practise the 8s

You should know all of the 8 times table now, but how quickly can you remember it?
Ask someone to time you as you do this page.
Remember, you must be fast but also correct!

1 x 8 = 8	8 x 10 = 80	8 x 6 = 48
2 x 8 = 16	2 x 8 = 16	3 x 8 = 24
3 x 8 = 24	4 x 8 = 32	9 x 8 = 72
4 x 8 = 32	6 x 8 = 48	8 x 4 = 32
5 x 8 = 40	8 x 8 = 64	11 x 8 = 88
6 x 8 = 48	12 x 8 = 96	8 x 2 = 16
7 x 8 = 56	1 x 8 = 8	7 x 8 = 56
8 x 8 = 64	3 x 8 = 24	12 x 8 = 96
9 x 8 = 72	5 x 8 = 40	8 x 3 = 24
10 x 8 = 80	7 x 8 = 56	5 x 8 = 40
11 x 8 = 88	8 x 3 = 24	8 x 8 = 64
12 x 8 = 96	8 x 5 = 40	2 x 8 = 16
8 x 2 = 16	8 x 8 = 64	8 x 9 = 72
8 x 3 = 24	8 x 9 = 72	4 x 8 = 32
8 x 4 = 32	8 x 11 = 88	8 x 7 = 56
8 x 5 = 40	8 x 4 = 32	10 x 8 = 80
8 x 6 = 48	8 x 6 = 48	8 x 12 = 96
8 x 7 = 56	8 x 8 = 64	8 x 0 = 0
8 x 8 = 64	8 x 10 = 80	8 x 11 = 88
8 x 9 = 72	8 x 0 = 0	12 x 8 = 96

Speed trials

You should know all of the 2, 3, 4, 5, 6, 7, 8, and 10 times tables now,
but how quickly can you remember them?
Ask someone to time you as you do this page.
Remember, you must be fast but also correct!

4 x 8 = 32	7 x 8 = 56	9 x 8 = 72
5 x 11 = 55	8 x 7 = 56	7 x 6 = 42
7 x 8 = 56	6 x 8 = 48	8 x 3 = 24
8 x 5 = 40	8 x 11 = 88	8 x 8 = 64
6 x 11 = 66	6 x 3 = 18	7 x 4 = 28
8 x 7 = 56	7 x 7 = 49	0 x 8 = 0
5 x 8 = 40	5 x 6 = 30	3 x 7 = 21
9 x 8 = 72	6 x 7 = 42	2 x 8 = 16
8 x 8 = 64	7 x 12 = 84	7 x 3 = 21
7 x 6 = 42	6 x 9 = 54	0 x 8 = 0
7 x 5 = 35	5 x 8 = 40	12 x 8 = 96
6 x 8 = 48	8 x 4 = 32	6 x 2 = 12
6 x 7 = 42	0 x 8 = 0	8 x 6 = 48
5 x 7 = 35	5 x 9 = 45	7 x 8 = 56
8 x 4 = 32	7 x 6 = 42	6 x 5 = 30
7 x 11 = 77	8 x 3 = 24	8 x 10 = 80
2 x 8 = 16	9 x 6 = 54	8 x 7 = 56
4 x 7 = 28	4 x 12 = 48	5 x 12 = 60
6 x 9 = 54	9 x 10 = 90	8 x 2 = 16
9 x 10 = 90	6 x 6 = 36	8 x 9 = 72

Some of the 9s

You should already know nearly all of the 9 times table because it is part of
the 2, 3, 4, 5, 6, 7, 8, and 10 times tables.
1 x 9 = 9   2 x 9 = 18   3 x 9 = 27   4 x 9 = 36   5 x 9 = 45
6 x 9 = 54   7 x 9 = 63   8 x 9 = 72   10 x 9 = 90
Find out if you can remember them quickly and correctly.

Cover the 9 times table with some paper so you can't see the numbers.
Write the answers as quickly as you can.

What are three nines?	27	What are ten nines?	90
What are two nines?	18	What are four nines?	36
What are six nines?	54	What are five nines?	45
What are seven nines?	63	What are eight nines?	72

Write the answers as quickly as you can.

How many nines are the same as 18?	2	How many nines are the same as 54?	6
How many nines are the same as 90?	10	How many nines are the same as 27?	3
How many nines are the same as 72?	8	How many nines are the same as 36?	4
How many nines are the same as 45?	5	How many nines are the same as 63?	7

Write the answers as quickly as you can.

Multiply nine by seven.	63	Multiply nine by ten.	90
Multiply nine by two.	18	Multiply nine by five.	45
Multiply nine by six.	54	Multiply nine by four.	36
Multiply nine by three.	27	Multiply nine by eight.	72

Write the answers as quickly as you can.

6 x 9 =	54	2 x 9 =	18	10 x 9 =	90
5 x 9 =	45	3 x 9 =	27	8 x 9 =	72
0 x 9 =	0	7 x 9 =	63	4 x 9 =	36

The rest of the 9s

You only need to learn these parts of the 9 times table.
9 x 9 = 81   9 x 11 = 99   9 x 12 = 108

This work will help you remember the 9 times table.

Complete these sequences.

9   18   27   36   45   54   63   72   81   90

8 x 9 = 72   so   9 x 9 = 72   plus another 9 =   81

45   54   63   72   81   90   99   108

9   18   27   36   45   54   63   72   81   90

9   18   27   36   45   54   63   72   81   90

Look for patterns in the 9 times table up to 10 x 9.

1	x	9	=	09
2	x	9	=	18
3	x	9	=	27
4	x	9	=	36
5	x	9	=	45
6	x	9	=	54
7	x	9	=	63
8	x	9	=	72
9	x	9	=	81
10	x	9	=	90

Write down any patterns you can see. There is more than one!

The digits in every answer add up to give 9.

If we take the first number of every answer, from top to bottom, we get

0, 1, 2, 3, 4, 5, 6, 7, 8, 9.

If we take the second number of every answer, from bottom to top, we get

0, 1, 2, 3, 4, 5, 6, 7, 8, 9, again

The first and the last answers are opposites (09 and 90), the second and the second last answers are opposites (18 and 81) and so on.

Practise the 9s

You should know all of the 9 times table now, but how quickly can you remember it? Ask someone to time you as you do this page. Remember, you must be fast but also correct!

1 x 9 =	9	9 x 10 =	90	9 x 6 =	54
2 x 9 =	18	2 x 9 =	18	3 x 9 =	27
3 x 9 =	27	4 x 9 =	36	9 x 9 =	81
4 x 9 =	36	6 x 9 =	54	9 x 4 =	36
5 x 9 =	45	9 x 7 =	63	11 x 9 =	99
6 x 9 =	54	17 x 9 =	108	9 x 2 =	18
7 x 9 =	63	1 x 9 =	9	7 x 9 =	63
8 x 9 =	72	3 x 9 =	27	12 x 9 =	108
9 x 9 =	81	5 x 9 =	45	9 x 3 =	27
10 x 9 =	90	7 x 9 =	63	5 x 9 =	45
11 x 9 =	99	9 x 9 =	81	9 x 9 =	81
12 x 9 =	108	9 x 11 =	99	2 x 9 =	18
9 x 2 =	18	9 x 5 =	45	8 x 9 =	72
9 x 3 =	27	0 x 9 =	0	4 x 9 =	36
9 x 4 =	36	9 x 1 =	9	9 x 7 =	63
9 x 5 =	45	9 x 2 =	18	10 x 9 =	90
9 x 6 =	54	9 x 4 =	36	9 x 5 =	45
9 x 7 =	63	9 x 6 =	54	9 x 0 =	0
9 x 8 =	72	9 x 8 =	72	9 x 11 =	99
9 x 9 =	81	9 x 12 =	108	12 x 9 =	108

Speed trials

You should know all of the times tables by now, but how quickly can you remember them? Ask someone to time you as you do this page. Remember, you must be fast but also correct!

6 x 8 =	48	4 x 8 =	32	8 x 12 =	96
9 x 12 =	108	9 x 8 =	72	7 x 9 =	63
5 x 8 =	40	6 x 6 =	36	8 x 5 =	40
7 x 5 =	35	8 x 9 =	72	8 x 7 =	56
6 x 4 =	24	6 x 4 =	24	7 x 4 =	28
8 x 8 =	64	7 x 3 =	21	4 x 9 =	36
5 x 11 =	55	5 x 9 =	45	6 x 7 =	42
9 x 8 =	72	6 x 8 =	48	4 x 6 =	24
8 x 3 =	24	7 x 7 =	49	7 x 8 =	56
7 x 7 =	49	6 x 9 =	54	6 x 9 =	54
9 x 5 =	45	7 x 8 =	56	11 x 8 =	88
4 x 8 =	32	8 x 4 =	32	6 x 5 =	30
6 x 7 =	42	0 x 9 =	0	8 x 8 =	64
2 x 9 =	18	10 x 12 =	120	7 x 6 =	42
8 x 4 =	32	7 x 6 =	42	6 x 8 =	48
7 x 12 =	84	8 x 7 =	56	9 x 10 =	90
2 x 8 =	16	9 x 6 =	54	8 x 4 =	32
4 x 7 =	28	8 x 6 =	48	7 x 11 =	77
6 x 9 =	54	11 x 9 =	99	5 x 8 =	40
9 x 9 =	81	6 x 7 =	42	8 x 9 =	72

Times tables for division

Knowing the times tables can also help with division sums.
Look at these examples.
3 x 6 = 18 which means that 18 ÷ 3 = 6 and that 18 ÷ 6 = 3
4 x 5 = 20 which means that 20 ÷ 4 = 5 and that 20 ÷ 5 = 4
9 x 11 = 99 which means that 99 ÷ 11 = 9 and that 99 ÷ 9 = 11

Use your knowledge of the times tables to work out these division sums.

3 x 8 = 24 which means that 24 ÷ 3 =   8   and that 24 ÷ 8 =   3
4 x 7 = 28 which means that 28 ÷ 4 =   7   and that 28 ÷ 7 =   4
3 x 5 = 15 which means that 15 ÷ 3 =   5   and that 15 ÷ 5 =   3
4 x 3 = 12 which means that 12 ÷ 3 =   4   and that 12 ÷ 4 =   3
3 x 11 = 33 which means that 33 ÷ 3 =   11   and that 33 ÷ 11 =   3
4 x 8 = 32 which means that 32 ÷ 4 =   8   and that 32 ÷ 8 =   4
3 x 9 = 27 which means that 27 ÷ 3 =   9   and that 27 ÷ 9 =   3
4 x 12 = 48 which means that 48 ÷ 4 =   12   and that 48 ÷ 12 =   4

These division sums help practise the 3 and 4 times tables.

20 ÷ 4 =	5	33 ÷ 3 =	11	16 ÷ 4 =	4
24 ÷ 4 =	6	27 ÷ 3 =	9	30 ÷ 3 =	10
12 ÷ 3 =	4	18 ÷ 3 =	6	28 ÷ 4 =	7
24 ÷ 3 =	8	48 ÷ 4 =	12	21 ÷ 3 =	7

How many fours in 36?	9	Divide 27 by three.	9
Divide 28 by 4.	7	How many threes in 21?	7
How many fives in 35?	7	Divide 40 by 5.	8
Divide 15 by 3.	5	How many eights in 48?	6

Times tables for division

This page will help you remember times tables by dividing by 2, 3, 4, 5, and 10.

$20 \div 5 = 4$ $18 \div 3 = 6$ $60 \div 5 = 12$

Complete the sums.

$44 \div 4 = 11$	$14 \div 2 = 7$	$32 \div 4 = 8$
$25 \div 5 = 5$	$21 \div 3 = 7$	$16 \div 4 = 4$
$24 \div 4 = 6$	$28 \div 4 = 7$	$12 \div 2 = 6$
$45 \div 5 = 9$	$60 \div 5 = 12$	$12 \div 3 = 4$
$10 \div 2 = 5$	$40 \div 10 = 4$	$12 \div 4 = 3$
$20 \div 10 = 2$	$20 \div 2 = 10$	$20 \div 2 = 10$
$6 \div 2 = 3$	$18 \div 3 = 6$	$20 \div 4 = 5$
$24 \div 3 = 8$	$32 \div 4 = 8$	$20 \div 5 = 4$
$30 \div 5 = 6$	$40 \div 5 = 8$	$20 \div 10 = 2$
$36 \div 3 = 12$	$33 \div 3 = 11$	$18 \div 2 = 9$
$40 \div 5 = 8$	$6 \div 2 = 3$	$18 \div 3 = 6$
$21 \div 3 = 7$	$15 \div 3 = 5$	$15 \div 3 = 5$
$14 \div 2 = 7$	$24 \div 4 = 6$	$15 \div 5 = 3$
$27 \div 3 = 9$	$15 \div 5 = 3$	$24 \div 3 = 8$
$48 \div 4 = 12$	$10 \div 10 = 1$	$24 \div 2 = 12$
$15 \div 5 = 3$	$4 \div 2 = 2$	$50 \div 5 = 10$
$15 \div 3 = 5$	$9 \div 3 = 3$	$55 \div 5 = 11$
$20 \div 5 = 4$	$4 \div 4 = 1$	$30 \div 3 = 10$
$20 \div 4 = 5$	$10 \div 5 = 2$	$30 \div 5 = 6$
$16 \div 2 = 8$	$110 \div 10 = 11$	$30 \div 10 = 3$

Times tables for division

This page will help you remember times tables by dividing by 2, 3, 4, 5, 6, 10, 11, and 12.

$30 \div 6 = 5$ $12 \div 6 = 2$ $66 \div 11 = 6$

Complete the sums.

$18 \div 6 = 3$	$27 \div 3 = 9$	$48 \div 6 = 8$
$30 \div 10 = 3$	$18 \div 6 = 3$	$35 \div 5 = 7$
$14 \div 2 = 7$	$22 \div 2 = 11$	$36 \div 4 = 9$
$18 \div 3 = 6$	$24 \div 6 = 4$	$24 \div 3 = 8$
$20 \div 4 = 5$	$24 \div 3 = 8$	$20 \div 2 = 10$
$15 \div 5 = 3$	$24 \div 4 = 6$	$33 \div 3 = 11$
$36 \div 6 = 6$	$30 \div 10 = 3$	$25 \div 5 = 5$
$55 \div 5 = 11$	$18 \div 2 = 9$	$32 \div 4 = 8$
$48 \div 4 = 12$	$18 \div 3 = 6$	$24 \div 2 = 12$
$15 \div 3 = 5$	$36 \div 4 = 9$	$16 \div 2 = 8$
$16 \div 4 = 4$	$36 \div 6 = 6$	$42 \div 6 = 7$
$25 \div 5 = 5$	$40 \div 5 = 8$	$5 \div 5 = 1$
$6 \div 6 = 1$	$120 \div 10 = 12$	$4 \div 4 = 1$
$10 \div 10 = 1$	$16 \div 4 = 4$	$28 \div 4 = 7$
$42 \div 6 = 7$	$12 \div 6 = 2$	$14 \div 2 = 7$
$24 \div 4 = 6$	$48 \div 12 = 4$	$24 \div 6 = 4$
$54 \div 6 = 9$	$54 \div 6 = 9$	$18 \div 6 = 3$
$99 \div 11 = 9$	$60 \div 6 = 10$	$54 \div 6 = 9$
$30 \div 6 = 5$	$66 \div 6 = 11$	$60 \div 6 = 10$
$30 \div 5 = 6$	$30 \div 6 = 5$	$40 \div 5 = 8$

Times tables for division

This page will help you remember times tables by dividing by 2, 3, 4, 5, 6, and 7.

$14 \div 7 = 2$ $28 \div 7 = 4$ $84 \div 7 = 12$

Complete the sums.

$21 \div 7 = 3$	$77 \div 7 = 11$	$84 \div 7 = 12$
$35 \div 5 = 7$	$28 \div 7 = 4$	$35 \div 5 = 7$
$14 \div 2 = 7$	$24 \div 6 = 4$	$35 \div 7 = 5$
$18 \div 6 = 3$	$24 \div 4 = 6$	$24 \div 6 = 4$
$20 \div 5 = 4$	$24 \div 2 = 12$	$21 \div 3 = 7$
$15 \div 3 = 5$	$21 \div 7 = 3$	$70 \div 7 = 10$
$36 \div 4 = 9$	$42 \div 7 = 6$	$42 \div 7 = 6$
$55 \div 5 = 11$	$18 \div 3 = 6$	$32 \div 4 = 8$
$18 \div 2 = 9$	$49 \div 7 = 7$	$27 \div 3 = 9$
$15 \div 5 = 3$	$36 \div 4 = 9$	$16 \div 4 = 4$
$48 \div 4 = 12$	$36 \div 3 = 12$	$42 \div 6 = 7$
$25 \div 5 = 5$	$40 \div 5 = 8$	$45 \div 5 = 9$
$7 \div 7 = 1$	$70 \div 7 = 10$	$84 \div 7 = 12$
$63 \div 7 = 9$	$24 \div 3 = 8$	$24 \div 3 = 8$
$42 \div 7 = 6$	$42 \div 6 = 7$	$14 \div 7 = 2$
$24 \div 2 = 12$	$48 \div 6 = 8$	$24 \div 4 = 6$
$54 \div 6 = 9$	$54 \div 6 = 9$	$18 \div 3 = 6$
$28 \div 7 = 4$	$60 \div 6 = 10$	$56 \div 7 = 8$
$30 \div 6 = 5$	$66 \div 6 = 11$	$63 \div 7 = 9$
$35 \div 7 = 5$	$25 \div 5 = 5$	$48 \div 6 = 8$

Times tables for division

This page will help you remember times tables by dividing by 2, 3, 4, 5, 6, 7, 8, and 9.

$16 \div 8 = 2$ $35 \div 7 = 5$ $27 \div 9 = 3$

$42 \div 6 = 7$	$81 \div 9 = 9$	$56 \div 7 = 8$
$32 \div 8 = 4$	$56 \div 7 = 8$	$45 \div 5 = 9$
$14 \div 7 = 2$	$63 \div 7 = 9$	$35 \div 7 = 5$
$48 \div 4 = 12$	$24 \div 6 = 4$	$18 \div 9 = 2$
$63 \div 7 = 9$	$22 \div 2 = 11$	$21 \div 3 = 7$
$72 \div 9 = 8$	$72 \div 9 = 8$	$28 \div 7 = 4$
$72 \div 8 = 9$	$42 \div 6 = 7$	$60 \div 5 = 12$
$56 \div 7 = 8$	$108 \div 9 = 12$	$32 \div 8 = 4$
$18 \div 6 = 3$	$14 \div 7 = 2$	$27 \div 9 = 3$
$81 \div 9 = 9$	$36 \div 4 = 9$	$16 \div 8 = 2$
$63 \div 9 = 7$	$36 \div 6 = 6$	$72 \div 6 = 12$
$45 \div 5 = 9$	$48 \div 8 = 6$	$45 \div 9 = 5$
$54 \div 9 = 6$	$21 \div 7 = 3$	$40 \div 4 = 10$
$70 \div 7 = 10$	$24 \div 3 = 8$	$24 \div 8 = 3$
$42 \div 7 = 6$	$40 \div 8 = 5$	$63 \div 7 = 9$
$30 \div 5 = 6$	$45 \div 9 = 5$	$24 \div 6 = 4$
$54 \div 6 = 9$	$54 \div 6 = 9$	$18 \div 6 = 3$
$56 \div 8 = 7$	$99 \div 9 = 11$	$96 \div 8 = 12$
$66 \div 6 = 11$	$63 \div 9 = 7$	$99 \div 9 = 11$
$35 \div 7 = 5$	$50 \div 5 = 10$	$48 \div 8 = 6$

Times tables practice grids

This is a times tables grid.

X	3	4	5
7	21	28	35
8	24	32	40

Complete each times tables grid.

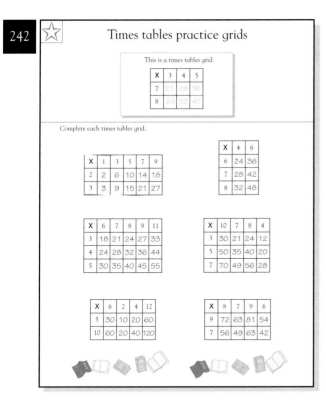

X	1	3	5	7	9
2	2	6	10	14	18
3	3	9	15	21	27

X	4	6
6	24	36
7	28	42
8	32	48

X	6	7	8	9	11
3	18	21	24	27	33
4	24	28	32	36	44
5	30	35	40	45	55

X	10	7	8	4
3	30	21	24	12
5	50	35	40	20
7	70	49	56	28

X	6	2	4	12
5	30	10	20	60
10	60	20	40	120

X	8	7	9	6
9	72	63	81	54
7	56	49	63	42

Times tables practice grids

Here are some more times tables grids.

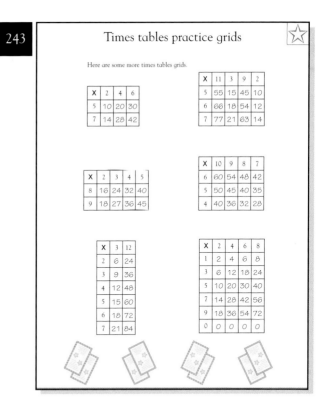

X	2	4	6
5	10	20	30
7	14	28	42

X	11	3	9	2
5	55	15	45	10
6	66	18	54	12
7	77	21	63	14

X	2	3	4	5
8	16	24	32	40
9	18	27	36	45

X	10	9	8	7
6	60	54	48	42
5	50	45	40	35
4	40	36	32	28

X	3	12
2	6	24
3	9	36
4	12	48
5	15	60
6	18	72
7	21	84

X	2	4	6	8
1	2	4	6	8
3	6	12	18	24
5	10	20	30	40
7	14	28	42	56
9	18	36	54	72
0	0	0	0	0

Times tables practise grids

Here are some more times tables grids.

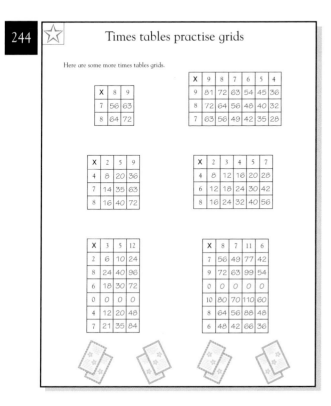

X	8	9
7	56	63
8	64	72

X	9	8	7	6	5	4
9	81	72	63	54	45	36
8	72	64	56	48	40	32
7	63	56	49	42	35	28

X	2	5	9
4	8	20	36
7	14	35	63
8	16	40	72

X	2	3	4	5	7
4	8	12	16	20	28
6	12	18	24	30	42
8	16	24	32	40	56

X	3	5	12
2	6	10	24
8	24	40	96
6	18	30	72
0	0	0	0
4	12	20	48
7	21	35	84

X	8	7	11	6
7	56	49	77	42
9	72	63	99	54
0	0	0	0	0
10	80	70	110	60
8	64	56	88	48
6	48	42	66	36

Speed trials

Try this final test.

27 ÷ 3 = 9	4 x 9 = 36	14 ÷ 2 = 7
7 x 9 = 63	18 ÷ 2 = 9	9 x 9 = 81
64 ÷ 8 = 8	6 x 8 = 48	15 ÷ 3 = 5
90 ÷ 10 = 9	21 ÷ 3 = 7	8 x 12 = 96
6 x 8 = 48	9 x 7 = 63	24 ÷ 3 = 8
45 ÷ 9 = 5	32 ÷ 4 = 8	7 x 8 = 56
3 x 12 = 36	4 x 11 = 44	30 ÷ 5 = 6
9 x 5 = 45	45 ÷ 5 = 9	6 x 6 = 36
48 ÷ 6 = 8	8 x 5 = 40	42 ÷ 6 = 7
7 x 7 = 49	42 ÷ 6 = 7	9 x 12 = 108
3 x 11 = 33	7 x 4 = 28	49 ÷ 7 = 7
56 ÷ 8 = 7	35 ÷ 7 = 5	8 x 6 = 48
36 ÷ 4 = 9	9 x 3 = 27	72 ÷ 8 = 9
24 ÷ 3 = 8	24 ÷ 8 = 3	9 x 7 = 63
36 ÷ 9 = 4	8 x 2 = 16	54 ÷ 9 = 6
6 x 7 = 42	36 ÷ 9 = 4	7 x 6 = 42
4 x 4 = 16	6 x 12 = 72	10 ÷ 10 = 1
32 ÷ 8 = 4	80 ÷ 10 = 8	7 x 11 = 77
49 ÷ 7 = 7	11 x 9 = 99	16 ÷ 8 = 2
25 ÷ 5 = 5	16 ÷ 2 = 8	7 x 9 = 63
56 ÷ 7 = 8	54 ÷ 9 = 6	63 ÷ 7 = 9

Practice Pages

Acknowledgements

Thank you to Alexander Cox, Wendy Horobin, Lorrie Mack and Penny Smith for editorial assistance.

Picture credits: The publisher would like to thank the following for their
kind permission to reproduce their photographs:
(Key: a-above; b-below/bottom; c-centre; f-far; l-left; r-right; t-top)

For Easy Peasy Times Tables:
CGTextures.com: 5 (balloon), 6ca, 6cr, 7 (all images), 8 (socks), 9-32 (all images);
Redrobes 6cla; Richard 6cl. **Corbis:** Image Source 8t (sky). **Fotolia:** Richard Blaker 6fcra.

For Tricky Times Tables:
DK Images: 62crb; Indianapolis Motor Speedway Foundation Inc. 36br; Lorraine Electronics Surveillance 58fbl,
63c; Natural History Museum, London 75cl; Stephen Oliver 39ca, 87ftr. **Getty Images:** Stone / Catherine Ledner
37cr. **iStockphoto.com:** Avava 37tl; bluestocking 36cb, 36clb, 36crb, 36fclb, 36fcrb; Joel Carillet 37tc; Angelo
Gilardelli 85cla; Lasavinaproduccions 71bl, 71cl, 71cla, 71clb; Vasko Miokovic 37tr; Skip ODonnell 37c, 79cr;
Denis Sarbashev 75tl; James Steidl 39cl, 39fcl; Ivonne Wierink-vanWetten 81tl.

All other images © Dorling Kindersley.
For further information see: www.dkimages.com